SENSITIVITY ANALYSIS IN PRACTICE

SENSITIVITY ANALYSIS IN PRACTICE
A GUIDE TO ASSESSING SCIENTIFIC MODELS

**Andrea Saltelli, Stefano Tarantola,
Francesca Campolongo and Marco Ratto**
Joint Research Centre of the European Commission, Ispra, Italy

John Wiley & Sons, Ltd

Other Wiley Editorial Offices

John Wiley & Sons Inc., 111 River Street, Hoboken, NJ 07030, USA

Jossey-Bass, 989 Market Street, San Francisco, CA 94103-1741, USA

Wiley-VCH Verlag GmbH, Boschstr. 12, D-69469 Weinheim, Germany

John Wiley & Sons Australia Ltd, 33 Park Road, Milton, Queensland 4064, Australia

John Wiley & Sons (Asia) Pte Ltd, 2 Clementi Loop #02-01, Jin Xing Distripark,
Singapore 129809

John Wiley & Sons Canada Ltd, 22 Worcester Road, Etobicoke, Ontario, Canada M9W 1L1

Wiley also publishes its books in a variety of electronic formats. Some content that appears
in print may not be available in electronic books.

Library of Congress Cataloging-in-Publication Data

Sensitivity analysis in practice : a guide to assessing scientific
models / Andrea Saltelli . . . [et al.].
      p.   cm.
   Includes bibliographical references and index.
   ISBN 0-470-87093-1   (cloth : alk. paper)
   1. Sensitivity theory (Mathematics)—Simulation methods.   2. SIMLAB.
I.   Saltelli, A.   (Andrea), 1953–
QA402.3   .S453   2004
003′ .5—dc22   2003021209

British Library Cataloguing in Publication Data

A catalogue record for this book is available from the British Library

ISBN 13: 978-0470-87093-8 (HB)

Typeset in 12/14pt Sabon by TechBooks, New Delhi, India

CONTENTS

PREFACE

This book is a 'primer' in global sensitivity analysis (SA). Its ambition is to enable the reader to apply global SA to a mathematical or computational model. It offers a description of a few selected techniques for sensitivity analysis, used for assessing the relative importance of model input factors. These techniques will answer questions of the type 'which of the uncertain input factors is more important in determining the uncertainty in the output of interest?' or 'if we could eliminate the uncertainty in one of the input factors, which factor should we choose to reduce the most the variance of the output?' Throughout this primer, the input factors of interest will be those that are uncertain, i.e. whose value lie within a finite interval of non-zero width. As a result, the reader will not find sensitivity analysis methods here that look at the local property of the input–output relationships, such as derivative-based analysis[1]. Special attention is paid to the selection of the method, to the framing of the analysis and to the interpretation and presentation of the results. The examples will help the reader to apply the methods in a way that is unambiguous and justifiable, so as to make the sensitivity analysis an added value to model-based studies or assessments. Both diagnostic and prognostic uses of models will be considered (a description of these is in Chapter 2), and Bayesian tools of analysis will be applied in conjunction with sensitivity analysis. When discussing sensitivity with respect to factors, we shall interpret the term 'factor' in a very broad sense: a factor is anything that can be changed in a model prior to its execution. This also includes structural or epistemic sources of uncertainty. To make an example, factors will be presented in applications that are in fact 'triggers', used to select one model structure versus another, one mesh size versus another, or altogether different conceptualisations of a system.

---

[1] A cursory exception is in Chapter 1.

Often, models use multi-dimensional uncertain parameters and/or input data to define the geographically distributed properties of a natural system. In such cases, a reduced set of scalar factors has to be identified in order to characterise the multi-dimensional uncertainty in a condensed, but exhaustive fashion. Factors will be sampled either from their prior distribution, or from their posterior distribution, if this is available. The main methods that we present in this primer are all related to one another and are the method of Morris for factors' screening and variance-based measures[2]. Also touched upon are Monte Carlo filtering in conjunction with either a variance based method or a simple two-sample test such as the Smirnov test. All methods used in this book are model-free, in the sense that their application does not rely on special assumptions on the behaviour of the model (such as linearity, monotonicity and additivity of the relationship between input factors and model output).

The reader is encouraged to replicate the test cases offered in this book before trying the methods on the model of interest. To this effect, the SIMLAB software for sensitivity analysis is offered. It is available free on the Web-page of this book http://www.jrc.cec.eu.int/uasa/primer-SA.asp. Also available at the same URL are a set of scripts in MATLAB® and the GLUEWIN software that implements a combination of global sensitivity analysis, Monte Carlo filtering and Bayesian uncertainty estimation.

This book is organised as follows. The first chapter presents the reader with most of the main concepts of the book, through their application to a simple example, and offers boxes with recipes to replicate the example using SIMLAB. All the concepts will then be revisited in the subsequent chapters. In Chapter 2 we offer another preview of the contents of the book, introducing succinctly the examples and their role in the primer. Chapter 2 also gives some definitions of the subject matter and ideas about the framing of the sensitivity analysis in relation to the defensibility of model-based assessment. Chapter 3 gives a full description of the test cases. Chapter 4 tackles screening methods for

---

[2] Variance based measures are generally estimated numerically using either the method of Sobol' or FAST (Fourier Analysis Sensitivity Test), or extensions of these methods available in the SIMLAB software that comes with this primer.

sensitivity analysis, and in particular the method of Morris, with applications. Chapter 5 discusses variance based measures, with applications. More ideas about 'setting for the analysis' are presented here. Chapter 6 covers Bayesian uncertainty estimation and Monte Carlo filtering, with emphasis on the links with global sensitivity analysis. Chapter 7 gives some instructions on how to use SIMLAB and, finally, Chapter 8 gives a few concepts and some opinions of various practitioners about SA and its implication for an epistemology of model use in the scientific discourse.

1 A WORKED EXAMPLE

This chapter presents an exhaustive analysis of a simple example, in order to give the reader a first overall view of the problems met in quantitative sensitivity analysis and the methods used to solve them. In the following chapters the same problems, questions, and techniques will be presented in full detail.

We start with a sensitivity analysis for a mathematical model in its simplest form, and work it out adding complications to it one at a time. By this process the reader will meet sensitivity analysis methods of increasing complexity, starting from the elementary approaches to the more quantitative ones.

1.1 A simple model

A simple portfolio model is:

$$Y = C_s P_s + C_t P_t + C_j P_j \tag{1.1}$$

where Y is the estimated risk[1] in €, C_s, C_t, C_j are the quantities per item, and P_s, P_t, P_j are *hedged* portfolios in €.[2] This means that each P_x, $x = \{s, t, j\}$ is composed of more than one item – so that the average return P_x is zero €. For instance, each hedged portfolio could be composed of an option plus a certain amount of underlying stock offsetting the option risk exposure due to

---

[1] This is the common use of the term. Y is in fact a return. A negative uncertain value of Y is what constitutes the risk.

[2] This simple model could well be seen as a composite (or synthetic) indicator camp by aggregating a set of standardised base indicators P_i with weights C_i (Tarantola *et al.*, 2002; Saisana and Tarantola, 2002).

Sensitivity Analysis in Practice: A Guide to Assessing Scientific Models A. Saltelli, S. Tarantola, F. Campolongo and M. Ratto © 2004 John Wiley & Sons, Ltd. ISBN 0-470-87093-1

movements in the market stock price. Initially we assume C_s, C_t, C_j = constants. We also assume that an estimation procedure has generated the following distributions for P_s, P_t, P_j:

$$
\begin{aligned}
P_s &\sim N(\bar{p}_s, \sigma_s), & \bar{p}_s &= 0, & \sigma_s &= 4 \\
P_t &\sim N(\bar{p}_t, \sigma_t), & \bar{p}_t &= 0, & \sigma_t &= 2 \\
P_j &\sim N(\bar{p}_j, \sigma_j), & \bar{p}_j &= 0, & \sigma_j &= 1.
\end{aligned}
\tag{1.2}
$$

The P_xs are assumed independent for the moment. As a result of these assumptions, Y will also be normally distributed with parameters

$$
\bar{y} = C_s\,\bar{p}_s + C_t\,\bar{p}_t + C_j\,\bar{p}_j \tag{1.3}
$$

$$
\sigma_y = \sqrt{C_s^2 \sigma_s^2 + C_t^2 \sigma_t^2 + C_j^2 \sigma_j^2}. \tag{1.4}
$$

Box 1.1   SIMLAB

The reader may want at this stage, or later in the study, to get started with SIMLAB by reproducing the results (1.3)–(1.4). This is in fact an uncertainty analysis, e.g. a characterisation of the output distribution of Y given the uncertainties in its input. The first thing to do is to input the factors P_s, P_t, P_j with the distributions given in (1.2). This is done using the left-most panel of SIMLAB (Figure 7.1), as follows:

1. Select 'New Sample Generation', then 'Configure', then 'Create New' when the new window 'STATISTICAL PRE PROCESSOR' is displayed.

2. Select 'Add' from the input factor selection panel and add factors one at a time as instructed by SIMLAB. Select 'Accept factors' when finished. This takes the reader back to the 'STATISTICAL PRE PROCESSOR' window.

3. Select a sampling method. Enter 'Random' to start with, and 'Specify switches' in the right. Enter something as a seed for random number generation and the number of executions (e.g. 1000). Create an output file by giving it a name and selecting a directory.

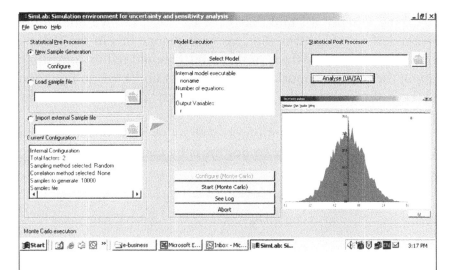

4. Go back to the left-most part of the SIMLAB main menu and click on 'Generate'. A sample is now available for the simulation.

5. We now move to the middle of the panel (Model execution) and select 'Configure (Monte Carlo)' and 'Select Model'. A new panel appears.

6. Select 'Internal Model' and 'Create new'. A formula parser appears. Enter the name of the output variable, e.g. 'Y' and follow the SIMLAB formula editor to enter Equation (1.1) with values of C_s, C_t, C_j of choice.

7. Select 'Start Monte Carlo' from the main model panel. The model is now executed the required number of times.

8. Move to the right-most panel of SIMLAB. Select 'Analyse UA/SA', select 'Y' as the output variable as prompted; choose the single time point option. This is to tell SIMLAB that in this case the output is not a time series.

9. Click on UA. The figure on this page is produced. Click on the square dot labelled 'Y' on the right of the figure and read the mean and standard deviation of Y. You can now compare these sample estimates with Equations (1.3–1.4).

Let us initially assume that $C_s < C_t < C_j$, i.e. we hold more of the less volatile items (but we shall change this in the following). A sensitivity analysis of this model should tell us something about the relative importance of the uncertain factors in Equation (1.1) in determining the output of interest Y, the risk from the portfolio.

According to first intuition, as well as to most of the existing literature on SA, the way to do this is by computing derivatives, i.e.

$$S_x^d = \frac{\partial Y}{\partial P_x}, \text{ with } x = s, t, j \qquad (1.5)$$

where the superscript 'd' has been added to remind us that this measure is in principle dimensioned ($\partial Y / \partial P_x$ is in fact dimensionless, but $\partial Y / \partial C_x$ would be in €). Computing S_x^d for our model we obtain

$$S_x^d = C_x, \text{ with } x = s, t, j. \qquad (1.6)$$

If we use the S_x^ds as our sensitivity measure, then the order of importance of our factors is $P_j > P_t > P_s$, based on the assumption $C_s < C_t < C_j$. S_x^d gives us the increase in the output of interest Y per unit increase in the factor P_x. There seems to be something wrong with this result: we have more items of portfolio j but this is the one with the least volatility (it has the smallest standard deviation, see Equation (1.2)). Even if $\sigma_s \gg \sigma_t, \sigma_j$, Equation (1.6) would still indicate P_j to be the most important factor, as Y would be locally more sensitive to it than to either P_t or P_s.

Sometime local sensitivity measures are normalised by some reference or central value. If

$$y^0 = C_s p_s^0 + C_t p_t^0 + C_j p_j^0. \qquad (1.7)$$

then one can compute

$$S_x^l = \frac{p_x^0}{y^0} \frac{\partial Y}{\partial P_x}, \text{ with } x = s, t, j. \qquad (1.8)$$

Applying this to our model, Equation (1.1), one obtains:

$$S_x^l = C_x \frac{p_x^0}{y^0}, \text{ with } x = s, t, j. \tag{1.9}$$

In this case the order of importance of the factors depends on the relative value of the C_xs weighted by the reference values p_x^0s. The superscript 'l' indicates that this index can be written as a logarithmic ratio if the derivative is computed in p_x^0.

$$S_x^l = \frac{p_x^0}{y^0} \frac{\partial Y}{\partial P_x}\bigg|_{y^0, p_x^0} = \frac{\partial \ln(Y)}{\partial \ln(P_x)}\bigg|_{y^0, p_x^0}. \tag{1.10}$$

S_x^l gives the fractional increase in Y corresponding to a unit fractional increase in P_x. Note that the reference point p_s^0, p_t^0, p_j^0 might be made to coincide with the vector of the mean values $\bar{p}_s, \bar{p}_t, \bar{p}_j$, though this would not in general guarantee that $\bar{y} = Y(\bar{p}_s, \bar{p}_t, \bar{p}_j)$, even though this is now the case (Equation (1.3)). Since $\bar{p}_s, \bar{p}_t, \bar{p}_j = 0$ and $\bar{y} = 0$, S_x^l collapses to be identical to S_x^d.

Also S_x^l is insensitive to the factors' standard deviations. It seems a better measure of importance than S_x^d, as it takes away the dimensions and is normalised, but it still offers little guidance as to how the uncertainty in Y depends upon the uncertainty in the P_xs.

A first step in the direction of characterising uncertainty is a normalisation of the derivatives by the factors' standard deviations:

$$S_s^\sigma = \frac{\sigma_s}{\sigma_y} \frac{\partial Y}{\partial P_s} = C_s \frac{\sigma_s}{\sigma_y}$$

$$S_t^\sigma = \frac{\sigma_t}{\sigma_y} \frac{\partial Y}{\partial P_t} = C_t \frac{\sigma_t}{\sigma_y} \tag{1.11}$$

$$S_j^\sigma = \frac{\sigma_j}{\sigma_y} \frac{\partial Y}{\partial P_j} = C_j \frac{\sigma_j}{\sigma_y}$$

where again the right-hand sides in (1.11) are obtained by applying Equation (1.1). Note that S_x^d and S_x^l are truly local in nature, as they

Table 1.1 S_x^σ measures for model (1.1) and different values of C_s, C_t, C_j (analytical values).

Factor	$C_s, C_t, C_j =$ 100, 500, 1000	$C_s, C_t, C_j =$ 300, 300, 300	$C_s, C_t, C_j =$ 500, 400, 100
P_s	0.272	0.873	0.928
P_t	0.680	0.436	0.371
P_j	0.680	0.218	0.046

need no assumption on the range of variation of a factor. They can be computed numerically by perturbing the factor around the base value. Sometimes they are computed directly from the solution of a differential equation, or by embedding sets of instructions into an existing computer program that computes Y. Conversely, S_x^σ needs assumptions to be made about the range of variation of the factor, so that although the derivative remains local in nature, S_x^σ is a hybrid local–global measure.

Also when using S_x^σ, the relative importance of P_s, P_t, P_j depends on the weights C_s, C_t, C_j (Table 1.1). An interesting result concerning the S_x^σs when applied to our portfolio model comes from the property of the model that $\sigma_y = \sqrt{C_s^2\sigma_s^2 + C_t^2\sigma_t^2 + C_j^2\sigma_j^2}$; squaring both sides and dividing by σ_y^2 we obtain

$$1 = \frac{C_s^2\sigma_s^2}{\sigma_y^2} + \frac{C_t^2\sigma_t^2}{\sigma_y^2} + \frac{C_j^2\sigma_j^2}{\sigma_y^2}. \tag{1.12}$$

Comparing (1.12) with (1.11) we see that for model (1.1) the squared S_x^σ give how much each individual factor contributes to the variance of the output of interest. If one is trying to assess how much the uncertainty in each of the input factors will affect the uncertainty in the model output Y, and if one accepts the variance of Y to be a good measure of this uncertainty, then the squared S_x^σ seem to be a good measure. However beware: the relation $1 = \sum_{x=s,t,j} (S_x^\sigma)^2$ is not general; it only holds for our nice, well hedged financial portfolio model. This means that you can still use S_x^σ if the input have a dependency structure (e.g. they are correlated) or the model is non-linear, but it is no longer true that the squared S_x^σ gives the exact fraction of variance attributable to each factor.

Using S_x^σ we see from Table 1.1 that for the case of equal weights ($= 300$), the factor that most influences the risk is the one with the highest volatility, P_s. This reconciles the sensitivity measure with our expectation.

Furthermore we can now put sensitivity analysis to use. For example, we can use the S_x^σ-based SA to build the portfolio (1.1) so that the risk Y is equally apportioned among the three items that compose it.

Let us now imagine that, in spite of the simplicity of the portfolio model, we chose to make a Monte Carlo experiment on it, generating a sample matrix

$$\mathbf{M} = \begin{matrix} p_s^{(1)} & p_t^{(1)} & p_j^{(1)} \\ p_s^{(2)} & p_t^{(2)} & p_j^{(2)} \\ \cdots & \cdots & \cdots \\ p_s^{(N)} & p_t^{(N)} & p_j^{(N)} \end{matrix} = [\mathbf{p}_s, \mathbf{p}_t, \mathbf{p}_j]. \tag{1.13}$$

$\mathbf{M}$ is composed of N rows, each row being a trial set for the evaluation of Y. The factors being independent, each column can be generated independently from the marginal distributions specified in (1.2) above. Computing Y for each row in $\mathbf{M}$ results in the output vector $\mathbf{y}$:

$$\mathbf{y} = \begin{matrix} y^{(1)} \\ y^{(2)} \\ \cdots \\ y^{(N)} \end{matrix} \tag{1.14}$$

An example of scatter plot (Y vs P_s) obtained with a Monte Carlo experiment of 1000 points is shown in Figure 1.1. Feeding both $\mathbf{M}$ and $\mathbf{y}$ into a statistical software (SIMLAB included), the analyst might then try a regression analysis for Y. This will return a model of the form

$$y^{(i)} = b_0 + b_s p_s^{(i)} + b_t p_t^{(i)} + b_j p_j^{(i)} \tag{1.15}$$

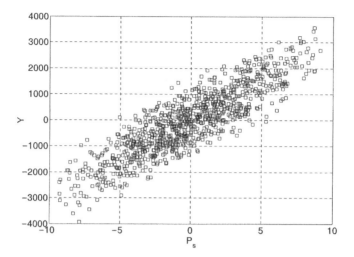

Figure 1.1   Scatter plot of Y vs. P_s for the model (1.1) $C_s = C_t = C_j = 300$. The scatter plot is made of $N = 1000$ points.

where the estimates of the b_xs are computed by the software based on ordinary least squares. Comparing (1.15) with (1.1) it is easy to see that if N is at least greater than 3, the number of factors, then $b_0 = 0$, $b_x = C_x$, $x = s, t, j$.

Normally one does not use the b_x coefficients for sensitivity analysis, as these are dimensioned. The practice is to computes the standardised regression coefficients (SRCs), defined as

$$\beta_x = b_x \sigma_x / \sigma_y. \tag{1.16}$$

These provide a regression model in terms of standardised variables

$$\tilde{\mathbf{y}} = \frac{\mathbf{y} - \bar{y}}{\sigma_y}; \quad \tilde{\mathbf{p}}_x = \frac{\mathbf{p}_x - \bar{p}_x}{\sigma_x} \tag{1.17}$$

i.e.

$$\tilde{\mathbf{y}} = \frac{\hat{\mathbf{y}} - \bar{y}}{\sigma_y} = \sum_{x=s,t,j} \beta_x \frac{\mathbf{p}_x - \bar{p}_x}{\sigma_x} = \sum_{x=s,t,j} \beta_x \tilde{\mathbf{p}}_x \tag{1.18}$$

where $\hat{y}$ is the vector of regression model predictions. Equation (1.16) tells us that the β_xs (standardised regression coefficients) for our portfolio model are equal to $C_x \sigma_x / \sigma_y$ and hence for linear models $\beta_x = S_x^\sigma$ because of (1.11). As a result, the values of the β_xs can also be read in Table 1.1.

Box 1.2   SIMLAB

You can now try out the relationship $\beta_x = S_x^\sigma$. If you have already performed all the steps in Box 1.1, you have to retrieve the saved input and output samples, so that you again reach step 9. Then:

10. On the right most part of the main SIMLAB panel, you activate the SA selection, and select SRC as the sensitivity analysis method.

11. You can now compare the SRC (i.e. the β_x) with the values in Table 1.1.

We can now try to generalise the results above as follows: for linear models composed of independent factors, the squared SRCs and S_x^σs provide the fraction of the variance of the model due to each factor.

For the standardised regression coefficients, these results can be further extended to the case of non-linear models as follows. The quality of regression can be judged by the model coefficient of determination R_y^2. This can be written as

$$R_y^2 = \frac{\sum_{i=1}^{N} (\hat{y}^{(i)} - \bar{y})^2}{\sum_{i=1}^{N} (y^{(i)} - \bar{y})^2} \tag{1.19}$$

where $\hat{y}^{(i)}$ is the regression model prediction. $R_y^2 \in [0, 1]$ represents the fraction of the model output variance accounted for by the regression model. The β_xs tell us how this fraction of the output variance can be decomposed according to the input factors, leaving us ignorant about the rest, where this rest is related to the non-linear part of the model. In the case of the linear model (1.1) we have, obviously, $R_y^2 = 1$.

The β_xs are a progress with respect to the S_x^σ; they can always be computed, also for non-linear models, or for models with no analytic representation (e.g. a computer program that computes Y). Furthermore the β_xs, unlike the S_x^σ, offer a measure of sensitivity that is multi-dimensionally averaged. While S_x^σ corresponds to a variation of factor x, all other factors being held constant, the β_x offers a measure of the effect of factor x that is averaged over a set of possible values of the other factors, e.g. our sample matrix (1.13). This does not make any difference for a linear model, but it does make quite a difference for non-linear models.

Given that it is fairly simple to compute standardised regression coefficients, and that decomposing the variance of the output of interest seems a sensible way of doing the analysis, why don't we always use the β_xs for our assessment of importance?

The answer is that we cannot, as often R_y^2 is too small, as e.g. in the case of non-monotonic models.[3]

1.2 Modulus version of the simple model

Imagine that the output of interest is no longer Y but its absolute value. This would mean, in the context of the example, that we want to study the deviation of our portfolio from risk neutrality. This is an example of a non-monotonic model, where the functional relationship between one (or more) input factor and the output is non-monotonic. For this model the SRC-based sensitivity analysis fails (see Box 1.3).

---

[3] Loosely speaking, the relationship between Y and an input factor X is monotonic if the curve $Y = f(X)$ is non-decreasing or non-increasing over all the interval of definition of X. A model with k factors is monotonic if the same rule applies for all factors. This is customarily verified, for numerical models, by Monte Carlo simulation followed by scatter-plots of Y versus each factor, one at a time.

Box 1.3   SIMLAB

Let us now estimate the coefficient of determination R_y^2 for the modulus version of the model.

1. Select 'Random sampling' with 1000 executions.

2. Select 'Internal Model' and click on the button 'Open existing configuration'. Select the internal model that you have previously created and click on 'Modify'.

3. The 'Internal Model' editor will appear. Select the formula and click on 'Modify'. Include the function 'fabs()' in the Expression editor. Accept the changes and go back to the main menu.

4. Select 'Start Monte Carlo' from the main model panel to generate the sample and execute the model.

5. Repeat the steps in Box 1.2 to see the results. The estimates of SRC appear with a red background as the test of significance is rejected. This means that the estimates are not reliable. The model coefficient of determination is almost null.

Is there a way to salvage our concept of decomposing the variance of Y into bits corresponding to the input factors, even for non-monotonic models? In general one has little a priori idea of how well behaved a model is, so that it would be handy to have a more robust variance decomposition strategy that works, whatever the degree of model non-monotonicity. These strategies are sometimes referred to as 'model free'.

One such strategy is in fact available, and fairly intuitive to get at. It starts with a simple question. If we could eliminate the uncertainty in one of the P_x, making it into a constant, how much would this reduce the variance of Y? Beware, for unpleasant models fixing a factor might actually increase the variance instead of reducing it! It depends upon where P_x is fixed.

The problem could be: how does $V_y = \sigma_y^2$ change if one can fix a generic factor P_x at its mid-point? This would be measured by $V(Y|P_x = \bar{p}_x)$. Note that the variance operator means in this case that while keeping, say, P_j fixed to the value $\bar{p}_j$ we integrate over P_s, P_t.

$$V(Y|\bar{P}_j = \bar{p}_j) = \int_{-\infty}^{+\infty}\int_{-\infty}^{+\infty} N(\bar{p}_s, \sigma_s)\, N(\bar{p}_t, \sigma_s)\,[(C_s P_s + C_t P_t + C_j \bar{p}_j)$$
$$- (C_s \bar{p}_s + C_t \bar{p}_t + C_j \bar{p}_j)]^2 \mathrm{d}P_s \mathrm{d}P_t. \qquad (1.20)$$

In practice, beside the problem already mentioned that $V(Y|P_x = \bar{p}_x)$ can be bigger than V_y, there is the practical problem that in most instances one does not know where a factor is best fixed. This value could be the true value, which is unknown at the simulation stage.

It sounds sensible then to average the above measure $V(Y|P_x = \bar{p}_x)$ over all possible values of P_x, obtaining $E(V(Y|P_x))$. Note that for the case, e.g. $x = j$, we could have written $E_j(V_{s,t}(Y|P_j))$ to make it clear that the average operator is over P_j and the variance operator is over P_s, P_t. Normally, for a model with k input factors, one writes $E(V(Y|X_j))$ with the understanding that V is over $\mathbf{X}_{-j}$ (a $(k-1)$ dimensional vector of all factors but X_j) and E is over X_j.

$E(V(Y|P_x))$ seems a good measure to use to decide how influential P_x is. The smaller the $E(V(Y|P_x))$, the more influential the factor P_x is. Textbook algebra tells us that

$$V_y = E(V(Y|P_x)) + V(E(Y|P_x)) \qquad (1.21)$$

i.e. the two operations complement the total unconditional variance. Usually $V(E(Y|P_x))$ is called the *main effect* of P_x on Y, and $E(V(Y|P_x))$ the *residual*. Given that $V(E(Y|P_x))$ is large if P_x is influential, its ratio to V_y is used as a measure of sensitivity, i.e.

$$S_x = \frac{V(E(Y|P_x))}{V_y} \qquad (1.22)$$

S_x is nicely scaled in [0, 1] and is variously called in the literature the *importance measure, sensitivity index, correlation ratio* or *first*

Table 1.2 S_x measures for model Y and different values of C_s, C_t, C_j (analytical values).

Factor	$C_s, C_t, C_j =$ 100, 500, 1000	$C_s, C_t, C_j =$ 300, 300, 300	$C_s, C_t, C_j =$ 500, 400, 100
P_s	0.074	0.762	0.860
P_t	0.463	0.190	0.138
P_j	0.463	0.048	0.002

order effect. It can be always computed, also for models that are not well-behaved, provided that the associate integrals exist. Indeed, if one has the patience to calculate the relative integrals in Equation (1.20) for our portfolio model, one will find that $S_x = (S_x^\sigma)^2 = \beta_x^2$, i.e. there is a one-to-one correspondence between the squared S_x^σ, the squared standardised regression coefficients and S_x for linear models with independent inputs. Hence all what we need to do to obtain the S_xs for the portfolio model (1.1) is to square the values in Table 1.1 (see Table 1.2). A nice property of the S_xs when applied to the portfolio model is that, for whatever combination of C_s, C_t, C_j, the sum of the three indices S_s, S_t, S_j is one, as one can easily verify (Table 1.2). This is not surprising, as the same was true for the β_x^2 when applied to our simple model. Yet the class of models for which this nice property of the S_xs holds is much wider (in practice that of the *additive models*[4]).

S_x is a good model-free sensitivity measure, and it always gives the expected reduction in the variance of the output that one would obtain if one could fix an individual factor.

As mentioned, for a system of k input uncertain factors, in general $\sum_{i=1}^{k} S_i \leq 1$.

Applying S_x to model $|Y|$, modulus of Y, one gets the estimations in Table 1.3. with SIMLAB.

We can see that the estimates of the expected reductions in the variance of $|Y|$ are much smaller than for Y. For example, in the case of $C_s, C_t, C_j = 300$, fixing P_s gives an expected variance reduction of 53% for $|Y|$, whilst the reduction of the variance for Y is 76%.

---

[4] A model $Y = f(X_1, X_2, \ldots, X_k)$ is additive if f can be decomposed as a sum of k functions f_i, each of which is a function only of the relative factor X_i.

Table 1.3   Estimation of S_xs for model $|Y|$ and different values of C_s, C_t, C_j.

Factor	$C_s, C_t, C_j =$ 100, 500, 1000	$C_s, C_t, C_j =$ 300, 300, 300	$C_s, C_t, C_j =$ 500, 400, 100
P_s	0	0.53	0.69
P_t	0.17	0.03	0.02
P_j	0.17	0	0

Given that the modulus version of the model is non-additive, the sum of the three indices S_s, S_t, S_j is less than one. For example, in the case $C_s, C_t, C_j = 300$, the sum is 0.56. What can we say about the remaining variance that is not captured by the S_xs? Let us answer this question not on the modulus version of model (1.1) but – for didactic purposes – on the slightly more complicated a six-factor version of our financial portfolio model.

Box 1.4   SIMLAB

Let us test the functioning of a variance-based technique with SIMLAB, by reproducing the results in Table 1.3

1. Select the 'FAST' sampling method and then 'Specify switches' on the right. Select 'Classic FAST' in the combo box 'Switch for FAST'. Enter something as seed and a number of executions (e.g. 1000). Create a sample file by giving it a name and selecting a directory.

2. Go back to the left-most part of the SIMLAB main menu and click on 'Generate'. A FAST-based sample is now available for the simulation.

3. Load the model with the absolute value as in Box 1.3 and click on 'Start (Monte Carlo)'.

4. Run the SA: a pie chart will appear reporting the estimated of S_x obtained with FAST. You can also see the tabulated values, which might be not as close to those reported in

Table 1.3 due to sampling error (the sample size is 1000). Try again with larger sample sizes and using the Sobol method, an alternative to the FAST method.

1.3 Six-factor version of the simple model

We now revert to model (1.1) and assume that the quantities C_xs are also uncertain. The model (1.1) now has six uncertain inputs. Let us assume

$$C_s \sim N(250, 200)$$
$$C_t \sim N(400, 300) \qquad (1.23)$$
$$C_j \sim N(500, 400).$$

The three distributions have been truncated at percentiles [11, 99.9], [10.0, 99.9] and [10.5, 99.9] respectively to ensure that $C_x > 0$.

There is no alternative now to a Monte Carlo simulation: the output distribution is in Figure 1.2, and the S_xs, as from Equation (1.22), are in Table 1.4. The S_xs have been estimated using a large

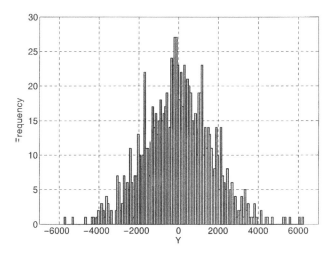

Figure 1.2 Output distribution for model (1.1) with six input factors, obtained from a Monte Carlo sample of 1000 elements.

Table 1.4 Estimates of first order effects S_x for model (1.1) with six input factors.

Factor	S_x
P_s	0.36
P_t	0.22
P_j	0.08
C_s	0.00
C_t	0.00
C_j	0.00
Sum	0.66

number of model evaluations (we will come back to this in future chapters; see also Box 1.5).

How is it that all effects for C_x are zero? All the P_x are centred on zero, and hence the conditional expectation value of Y is zero regardless of the value of C_x, i.e. for model (1.1) we have:

$$E(Y|C_x = c_x^*) = E(Y) = 0, \quad \text{for all } c_x^* \tag{1.24}$$

and as a result, $V(E(Y|C_x)) = 0$. This can also be visualised in Figure 1.3; inner conditional expectations of Y can be taken averaging along vertical 'slices' of the scatter plot. In the case of Y vs. C_s (lower panel) it is clear that such averages will form a perfectly horizontal line on the abscissas, implying a zero variance for the averaged Ys and a null sensitivity index. Conversely, for Y vs. P_s (upper panel) the averages along the vertical slices will form an increasing line, implying non-zero variance for the averaged Ys and a non-null sensitivity index.

As anticipated the S_xs do not add up to one. Let us now try a little experiment. Take two factors, say P_s, P_t, and estimate our sensitivity measure on the pair i.e. compute $V(E(Y|P_s, P_t))/V_y$. By definition this implies taking the average over all factors except P_s, P_t, and the variance over P_s, P_t. We do this (we will show how later) and call the results $S_{P_s P_t}^c$, where the reason for the superscript c will be clear in a moment. We see that $S_{P_s P_t}^c = 0.58$, i.e.

$$S_{P_s P_t}^c = \frac{V(E(Y|P_s, P_t))}{V_y} = S_{P_s} + S_{P_t} = 0.36 + 0.22. \tag{1.25}$$

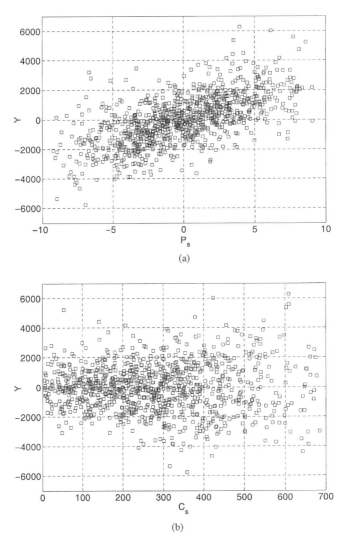

Figure 1.3   Scatter plots for model (1.1): (a) of Y vs. P_s, (b) of Y vs. C_s. The scatter plots are made up of $N = 1000$ points.

This seems a nice result. Let us try the same game with C_s, P_s. The results show that now:

$$S^c_{C_s P_s} = \frac{V(E(Y|C_s, P_s))}{V_y} = 0.54 > S_{C_s} + S_{P_s} = 0.36. \quad (1.26)$$

Table 1.5   Incomplete list of pair-wise effects S_{xz} for model (1.1) with six input factors.

Factors	S_{xz}
P_s, C_s	0.18
P_t, C_t	0.11
P_j, C_j	0.05
P_s, C_t	0.00
P_s, C_j	0.00
P_t, C_s	0.00
P_t, C_j	0.00
P_j, C_s	0.00
P_j, C_t	0.00
Sum of first order terms (Table 1.4)	0.66
Grand sum	1

Let us call $S_{C_s P_s}$ the difference

$$S_{C_s P_s} = \frac{V(E(Y|C_s, P_s))}{V_y} - \frac{V(E(Y|C_s))}{V_y} - \frac{V(E(Y|P_s))}{V_y}$$

$$= S_{C_s P_s}^c - S_{C_s} - S_{P_s}. \tag{1.27}$$

Values of this measure for pairs of factors are given in Table 1.5.

Note that we have not listed all effects of the type P_x, P_y, and C_x, C_y in this table as they are all null.

Trying to make sense of this result, one might ponder that if the combined effect of two factors, i.e. $V(E(Y|C_s, P_s))/V_y$, is greater than the sum of the individual effects $V(E(Y|C_s))/V_y$ and $V(E(Y|P_s))/V_y$, perhaps this extra variance describes a synergistic or co-operative effect between these two factors. This is in fact the case and $S_{C_s P_s}$ is called the *interaction* (or two-way) effect of C_s, P_s on Y and measures that part of the effect of C_s, P_s on Y that exceeds the sum of the first-order effects. The reason for the superscript c in $S_{C_s P_s}^c$ can now be explained: this means that the effect measured by $S_{C_s P_s}^c$ is *closed* over the factors C_s, P_s, i.e. by it we capture all the effects that include only these two factors. Clearly if there were a non-zero interaction between C_s and a third factor, say C_t, this would not be captured by $S_{C_s P_s}^c$. We see from

Tables 1.4–5 that if we sum all first-order with all second-order effects we indeed obtain 1, i.e. all the variance of Y is accounted for.

This is clearly only valid for our financial portfolio model because it only has interaction effects up to the second order; if we were to compute higher order effects, e.g.

$$S_{C_s P_s P_t} = S^c_{C_s P_s P_t} - S_{C_s P_s} - S_{P_s P_t} - S_{C_s P_t} - S_{C_s} - S_{P_s} - S_{P_t} \quad (1.28)$$

they would all be zero, as one may easily realise by inspecting Equation (1.1). $S^c_{C_s P_s P_t}$ on the other hand is non-zero, and is equal to the sum of the three second-order terms (of which only one differs from zero) plus the sum of three first-order effects. Specifically

$$S^c_{C_s P_s P_t} = 0.17 + 0.02 + 0.35 + 0.22 = 0.76.$$

The full story for these partial variances is that for a system with k factors there may be interaction terms up to the order k, i.e.

$$\sum_i S_i + \sum_i \sum_{j>i} S_{ij} + \sum_i \sum_{j>i} \sum_{l>j} S_{ijl} + \ldots S_{12\ldots k} = 1 \quad (1.29)$$

For the portfolio model with $k = 6$ all terms above the second order are zero and only three second-order terms are nonzero.

This is lucky, one might remark, because these terms would be a bit too numerous to look at. How many would there be? Six first order, $\binom{6}{2} = 15$ second order, $\binom{6}{3} = 20$ third order, $\binom{6}{4} = 15$ fourth order, $\binom{6}{5} = 6$ fifth order, and one, the last, of order $k = 6$. This makes 63, just equal to $2^k - 1 = 2^6 - 1$, which is the formula to use. This result seems to suggest that the S_i, and their higher order relatives S_{ij}, S_{ijl} are nice, informative and model free, but they may become cumbersomely too many for practical use unless the development (1.29) quickly converges to one. Is there a recipe for treating models that do not behave so nicely?

For this we use the so-called total effect terms, whose description is given next.

Let us go back to our portfolio model and call $\mathbf{X}$ the set of all factors, i.e.

$$\mathbf{X} \equiv (C_s, C_t, C_j, P_s, P_t, P_j) \tag{1.30}$$

and imagine that we compute:

$$\frac{V(E(Y|\mathbf{X}_{-C_s}))}{V_y} = \frac{V(E(Y|C_t, C_j, P_s, P_t, P_j))}{V_y} \tag{1.31}$$

(the all-but-C_s notation has been used). It should now be apparent that Equation (1.31) includes all terms in the development (1.29), of any order, that do not contain the factor C_s. Now what happens if we take the difference

$$1 - \frac{V(E(Y|\mathbf{X}_{-C_s}))}{V_y}? \tag{1.32}$$

The result is nice; for our model, where only a few higher-order terms are non-zero, it is

$$1 - \frac{V(E(Y|\mathbf{X}_{-C_s}))}{V_y} = S_{C_s} + S_{C_s P_s} \tag{1.33}$$

i.e. the sum of all non-zero terms that include C_s. The generalisation to a system with k factors is straightforward:

$$1 - \frac{V(E(Y|\mathbf{X}_{-i}))}{V_y} = \text{sum of all terms of any order that include the factor } X_i.$$

Note that because of Equation (1.21), $1 - V(E(Y|\mathbf{X}_{-i}))/V_y = E(V(Y|\mathbf{X}_{-i}))/V_y$. We indicate this as S_{Ti} and call it the total effect term for factor X_i. If we had computed the S_{Ti} indices for a three-factor model with orthogonal inputs, e.g. our modulus model of Section 1.2, we would have obtained, for example, for factor P_s:

$$S_{TP_s} = S_{P_s} + S_{P_s P_t} + S_{P_s P_j} + S_{P_s P_t P_j} \tag{1.34}$$

and similar formulae for P_t, P_j. For the modulus model, all terms in (1.34) could be non-zero. Another way of looking at the measures $V(E(Y|X_i))$, $E(V(Y|\mathbf{X}_{-i}))$ and the corresponding indices S_i, S_{Ti} is in terms of top- and bottom-marginal variances. We have already said that $E(V(Y|X_i))$ is the average output

Table 1.6   Estimates of the main effects and total effect indices for model (1.1) with six input factors.

Factor	S_x	S_{Tx}
P_s	0.36	0.57
P_t	0.22	0.35
P_j	0.08	0.14
C_s	0.00	0.19
C_t	0.00	0.12
C_j	0.00	0.06
Sum	0.66	1.43

variance that would be left if X_i could be known or could be fixed. Consequently $V(E(Y|X_i))$ is the expected reduction in the output variance that one would get if X_i could be known or fixed. Michiel J. W. Jansen, a Dutch statistician, calls this latter a top marginal variance. By definition the total effect measure $E(V(Y|\mathbf{X}_{-i}))$ is the expected residual output variance that one would end up with if all factors but X_i could be known or fixed. Hence the term, still due to Jansen, of bottom marginal variance. For the case of independent input variables, it is always true that $S_i \leq S_{Ti}$, where the equality holds for a purely additive model.

In a series of works published since 1993, we have argued that if one can compute all the k S_i terms plus all the k S_{Ti} ones, then one can obtain a fairly complete and parsimonious description of the model in terms of its global sensitivity analysis properties. The estimates for our six-factor portfolio model are given in Table 1.6.

As one might expect, the sum of the first-order terms is less than one, the sum of the total order effects is greater than one.

Box 1.5   SIMLAB

Let us try to obtain the numbers in Table 1.6 using SIMLAB. Remember to configure the set of factors so as to include the three factors C_x with their respective distributions and truncations (see Equations (1.23)).

1. Select the 'FAST' sampling method and then 'Specify switches' on the right. Now select 'All first and total order effect calculation (Extended FAST)' in the combo box 'Switch for FAST'. Enter an integer number as seed and the cost of the analysis in terms of number of model executions (e.g. 10 000).

2. Go back to the SIMLAB main menu and click on 'Generate'. After a few moments a sample is available for the simulation.

3. Load the model as in Box 1.3 and click on 'Start (Monte Carlo)'.

4. Run the SA: two pie charts will appear reporting both the S_x and S_{Tx} estimated with the Extended FAST. You can also look at the tabulated values. Try again using the method of Sobol'.

Here we anticipate that the cost of the analysis leading to Table 1.6 is $N(k + 2)$, where the cost is expressed in number of model evaluations and N is the column dimension of the Monte Carlo matrix used in the computations, say $N = 500$ to give an order of magnitude (in Box 1.5 $N = 1000/8 = 1250$). Computing all terms in the development (1.29) is more expensive, and often prohibitively so.[5] We would also anticipate, this time from Chapter 4, that a gross estimate of the S_{Tx} terms can be obtained at a lower cost using an extended version of the method of Morris. Also for this method the size is proportional to the number of factors.

1.4 The simple model 'by groups'

Is there a way to compact the results of the analysis further? One might wonder if one can get some information about the overall sensitivity pattern of our portfolio model at a lower price. In fact a nice property of the variance-based methods is that the variance

---

[5] The cost would be exponential in k, see Chapter 5.

Table 1.7 Estimates of main effects and total effect indices of two groups of factors of model (1.1).

Factor	S_x	S_{Tx}
$\mathbf{P} \equiv (P_s, P_t, P_j)$	0.66	1.00
$\mathbf{C} \equiv (C_s, C_t, C_j)$	0.00	0.34
Sum	0.66	1.34

decomposition (1.29) can be written for sets of factors as well. In our model, for instance, it would be fairly natural to write a variance decomposition as:

$$S_C + S_P + S_{C,P} = 1 \qquad (1.35)$$

where $\mathbf{C} = C_s, C_t, C_j$ and $\mathbf{P} = P_s, P_t, P_j$. The information we obtain in this way is clearly less than that provided by the table with all S_i and S_{Ti}.[6]

Looking at Table 1.7 we again see that the effect of the C set at the first order is zero, while the second-order term $S_{C,P}$ is 0.34, so it is not surprising that the sum of the total effects is 1.34 (the 0.34 is counted twice):

$$\begin{aligned} S_{TC} &= S_C + S_{C,P} \\ S_{TP} &= S_P + S_{C,P} \end{aligned} \qquad (1.36)$$

Now all that we know is the combined effect of all the amounts of hedges purchased, C_x, the combined effect of all the hedged portfolios, P_x, plus the interaction term between the two. Computing all terms in Equation 1.35 (Table 1.7) only costs $N \times 3$, one set of size N to compute the unconditional mean and variance, one for C and one for P, $S_{C,P}$ being computed by difference using (1.35). This is less than the $N \times (6 + 2)$ that one would have needed to compute all terms in Table 1.6. So there is less information at less cost, although cost might not be the only factor leading one to decide to present the results of a sensitivity analysis by groups. For instance, we could have shown the results from the portfolio model as

$$S_s + S_t + S_j + S_{s,t} + S_{t,j} + S_{s,j} + S_{s,t,j} = 1 \qquad (1.37)$$

---

[6] The first-order sensitivity index of a group of factors is equivalent to the closed effect of all the factors in the group, e.g.: $S_C = S^c_{C_s, C_t, C_j}$.

Table 1.8   Main effects and total effect indices of three groups of factors of model (1.1).

Factor	S_x	S_{Tx}
$s \equiv (C_s, P_s)$	0.54	0.54
$t \equiv (C_t, P_t)$	0.33	0.33
$j \equiv (C_j, P_j)$	0.13	0.13
Sum	1	1

where $s \equiv (C_s, P_s)$ and so on for each sub-portfolio item, where a sub-portfolio is represented by a certain amount of a given type of hedge. This time the problem has become additive, i.e. all terms of second and third order in (1.37) are zero. Given that the interactions are 'within' the groups of factors, the sum of the first-order effects for the groups is one, i.e. $S_s + S_t + S_j = 1$, and the total indices are the same as the main effect indices (Table 1.8).

Different ways of grouping the factors might give different insights into the owner of the problem.

Box 1.6   SIMLAB

Let us estimate the indices in Table 1.7 with SIMLAB.

1. Select the 'FAST' sampling method and then 'Specify switches' on the right. Now select 'All first and total order effect calculation on groups' in the combo box 'Switch for FAST'. Enter something as seed and a number of executions (e.g. 10 000).

2. Instead of generating the sample now, load the model first by clicking on 'Configure (Monte Carlo)' and then 'Select Model'.

3. Now click on 'Start (Monte Carlo)'. SIMLAB will generate the sample and run the model all together.

4. Run the SA: two pie charts will appear showing both the S_x and S_{Tx} estimated for the groups in Table 1.7. You can also look at the tabulated values. Try again using larger sample sizes.

1.5   The (less) simple correlated-input model

We have now reached a crucial point in our presentation. We have to abandon the last nicety of the portfolio model: the orthogonality (independence) of its input factors.[7]

We do this with little enthusiasm because the case of dependent factors introduces the following considerable complications.

1. Development (1.29) no longer holds, nor can any higher-order term be decomposed into terms of lower dimensionality, i.e. it is no longer true that

$$\frac{V(E(Y|C_s, P_s))}{V_y} = S^c_{C_s P_s} = S_{C_s} + S_{P_s} + S_{C_s P_s} \qquad (1.38)$$

although the left-hand side of this equation can be computed, as we shall show. This also impacts on our capacity to treat factors into sets, unless the non-zero correlations stay confined within sets, and not across them.

2. The computational cost increases considerably, as the Monte Carlo tricks used for non-orthogonal input are not as efficient as those for the orthogonal one.

Assume a non-diagonal covariance structure C for our problem:

$$
C = \begin{array}{c|cccccc}
 & P_s & P_t & P_j & C_s & C_t & C_j \\
\hline
P_s & 1 & & & & & \\
P_t & 0.3 & 1 & & & & \\
P_j & 0.3 & 0.3 & 1 & & & \\
C_s & . & . & . & 1 & & \\
C_t & . & . & . & -0.3 & 1 & \\
C_j & . & . & . & -0.3 & -0.3 & 1 \\
\end{array} \qquad (1.39)
$$

We assume the hedges to be positively correlated among one another, as each hedge depends upon the behaviour of a given stock

---

[7] The most intuitive type of dependency among input factors is given by correlation. However, dependency is a more general concept than correlation, i.e. independency means orthogonality and also implies that the correlation is null, while the converse is not true, i.e. null correlation does not necessarily imply orthogonality (see, for example, Figure 6.6 and the comments to it). The equivalence between null correlation and independency holds for multivariate normal distributions.

Table 1.9   Estimated main effects and total effect indices for model (1.1) with correlated inputs (six factors).

Factor	S_x	S_{Tx}
P_s	0.58	0.35
P_t	0.48	0.21
P_j	0.36	0.085
C_s	0.01	0.075
C_t	0.00	0.045
C_j	0.00	0.02
Sum	1.44	0.785

price and we expect the market price dynamics of different stocks to be positively correlated. Furthermore, we made the assumption that the C_xs are negatively correlated, i.e. when purchasing more of a given hedge investors tends to reduce their expenditure on another item.

The marginal distributions are still given by (1.2), (1.23) above. The main effect coefficients are given in Table 1.9. We have also estimated the S_{Tx} indices as, for example, for P_s:

$$S_{TP_s} = \frac{E(V(Y|\mathbf{X}_{-P_s}))}{V(Y)} = 1 - \frac{V(E(Y|\mathbf{X}_{-P_s}))}{V(Y)} \qquad (1.40)$$

with a brute force method at large sample size. The calculation of total indices for correlated input is not implemented in SIMLAB.

We see that now the total effect terms can be smaller than the first-order terms. This should be intuitive in terms of bottom marginal variances. Remember that $E(V(Y|\mathbf{X}_{-i}))$ is the expected residual variance that one would end up with if all factors but X_i could be known or fixed. Even if factor X_i is still non-determined, all other factors have been fixed, and on average one would be left with a smaller variance, than one would get for the orthogonal case, due to the relation between the fixed factors and the unfixed one. The overall result for a non-additive model with non-orthogonal inputs will depend on the relative predominance of the interaction, pushing for $S_{Ti} > S_i$ as for the C_xs in Table 1.9, and dependency between input factors, pushing for $S_{Ti} < S_i$ as for the P_xs in Table 1.9. For all additive models with non-orthogonal

Table 1.10 Decomposition of $V(Y)$ and relative value of $V(E(Y|X_i))$, $E(V(Y|\mathbf{X}_{-i}))$ for two cases: (1) orthogonal input, all models and (2) non-orthogonal input, additive models. When the input is non-orthogonal and the model non-additive, $V(E(Y|X_i))$ can be higher or lower than $E(V(Y|\mathbf{X}_{-i}))$.

Case (1) Orthogonal input factors, all models. For additive models the two rows are equal.	$V(E(Y|X_i))$ top marginal. (or main effect) of X_i	$E(V(Y|X_i))$ bottom marginal. (or total effect) of $\mathbf{X}_{-i}$
	$E(V(Y|\mathbf{X}_{-i}))$ bottom marginal (or total effect) of X_i	$V(E(Y|\mathbf{X}_{-i}))$ top marginal (or main effect) of $\mathbf{X}_{-i}$
Case (2) Non-orthogonal input factors, additive models only. If the dependency between inputs vanishes, the two rows become equal. For the case where X_i and $\mathbf{X}_{-i}$ are perfectly correlated both the $E(V(Y|.))$ disappear and both the $V(E(Y|.))$ become equal to $V(Y)$.	$V(E(Y|X_i))$ top marginal of X_i	$E(V(Y|X_i))$ bottom marginal of $\mathbf{X}_{-i}$
	$E(V(Y|\mathbf{X}_{-i}))$ bottom marginal of X_i	$V(E(Y|\mathbf{X}_{-i}))$ top marginal of $\mathbf{X}_{-i}$
	$V(Y)$ (Unconditional)	

input factors it will be $S_{Ti} \leq S_i$. In the absence of interactions (additive model) it will be $S_{Ti} = S_i$ for the orthogonal case. If, still with an additive model, we now start imposing a dependency among the input factors (e.g. adding a correlation structure), then S_{Ti} will start decreasing as $E(V(Y|\mathbf{X}_{-i}))$ will be lower because having conditioned on $\mathbf{X}_{-i}$ also limits the variation of X_i (Table 1.10).

We take factor P_s as an example for the discussion that follows. Given that for the correlated input case S_{TP_s} can no longer be thought of as the sum of all terms including factor P_s, what is the point of computing it? The answer lies in one of the possible uses that is made of sensitivity analysis: that of ascertaining if a given factor is so non-influential on the output (in terms of contribution to the output's variance as usual!) that we can fix it. We submit that if we want to fix a factor or group of factors, it is their S_{Tx} or S_{Tx} respectively that we have to look at. Imagine that we want to determine if the influence of P_s is zero. If P_s is totally non-influential, then surely

$$E(V(Y|\mathbf{X}_{-P_s})) = 0 \qquad\qquad (1.41)$$

because fixing 'all but P_s' results in the inner variance over P_s being zero (under that hypothesis the variance of Y is driven only by non-P_s), and this remains zero if we take the average over all possible values of non-P_s. As a result, S_{TP_s} is zero if P_s is totally non-influential. It is easy to see that the condition $E(V(Y|\mathbf{X}_{-P_s})) = 0$ is necessary and sufficient for factor P_s to be non-influent, under any model or correlation/dependency structure among input factors.

1.6 Conclusions

This ends our analysis of the model in Equation (1.1). Although we haven't given the reader any of the computational strategies to compute the S_x, S_{Tx} measures, it is easy to understand how these can be computed in principle. After all, a variance is an integral. It should be clear that under the assumptions that:

1. the model is not so terribly expensive that one cannot afford Monte Carlo simulations, and

2. one has a scalar objective function Y and is happy with its variance being the descriptor of interest.

Then

1. variance based measures offer a coherent strategy for the decomposition of the variance of Y;

2. the strategy is agile in that the owner of the problem can decide if and how to group the factors for the analysis;

3. this strategy is model free, i.e. it also works for nasty, non-monotonic, non-additive models Y, and converge to easy-to-grasp statistics, such as the squared standardised regression coefficients β_x^2 for the linear model;

4. it remains meaningful for the case where the input factors are non-orthogonal;

5. it lends itself to intuitive interpretations of the analysis, such as that in terms of top and bottom marginal variances, in terms of prioritising resources to reduce the uncertainty of the most influential factors or in terms of fixing non-influential factors.

2 GLOBAL SENSITIVITY ANALYSIS FOR IMPORTANCE ASSESSMENT

In this chapter we introduce some examples, most of which will later serve as test cases. The examples are described in detail in Chapter 3. Here a flash description is offered, for the purpose to illustrate different problem settings for SA. The hurried reader can use these descriptions to match the example with their own application. Next, a definition of sensitivity analysis is offered, complemented by a discussion of the desirable properties that a sensitivity analysis method should have. The chapter ends with a categorisation (a taxonomy) of application settings, that will help us to tackle the applications effectively and unambiguously.

2.1 Examples at a glance

Bungee jumping
We are physicist and decide to join a bungee-jumping club, but want to model the system first. The asphalt is at a distance H (not well quantified) from the launch platform. Our mass M is also uncertain and the challenge is to choose the best bungee cord ($\sigma =$ number of strands) that will allow us to almost touch the ground below, thus giving us real excitement. We do not want to use cords that give you a short (and not exciting) ride. So, we choose the minimum distance to the asphalt during the oscillation (h_{min}) as a convenient indicator of excitement. This indicator is a function of three variables: H, M and σ.

Our final target is to identify the best combination of the three variables that gives as the minimum value of h_{min} (though this must

Sensitivity Analysis in Practice: A Guide to Assessing Scientific Models A. Saltelli, S. Tarantola, F. Campolongo and M. Ratto © 2004 John Wiley & Sons, Ltd. ISBN 0-470-87093-1

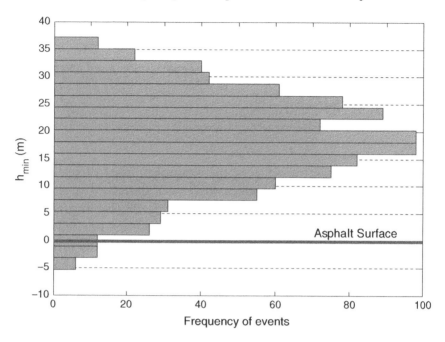

Figure 2.1 Uncertainty analysis of the bungee-jumping excitement indicator $h_{\min}$.

remain positive!). This search is a simple minimisation problem with constraints, which can be managed with standard optimisation techniques. However, the uncertanity in the three variables implies an uncertinity on $h_{\min}$. If such uncertainty is too wide, the risk of a failure in the jump is high and must be reduced. Under these circumstances, it is wise to investigate, through sensitivity analysis, which variable drives most of the uncertainty on $h_{\min}$. This indicates where one should improve our knowledge in order to reduce the risk of failure.

The uncertainty analysis (Figure 2.1), shows the uncertainty on $h_{\min}$ due to the uncertainties in H, M and σ (more details will be given in Chapter 3). The probability of a successful jump is 97.4%. The sensitivity analysis (Chapter 5) shows that the number of strands in the cord is the risk-governing variable, worth examining in more detail. Meanwhile, we should not waste time improving the knowledge of our mass M, as its effect on the uncertainty of $h_{\min}$ is very small.

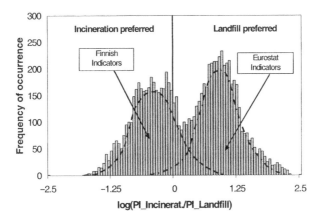

Figure 2.2 Monte Carlo based uncertainty analysis of $Y = $ Log $(PI$ (Incineration)/PI(Landfill)) obtained by propagating the uncertainty of waste inventories, emission factors, weights for the indicators etc. The uncertainty distribution is bimodal, with one maximum for each for the two waste management alternatives, making the issue non-decidable.

Decision analysis

A model is used to decide whether solid waste should be burned or disposed of in landfill, in an analysis based on Austrian data for 1994. This is a methodological exercise, described in detail in Saltelli *et al.* (2000a, p. 385). An hypothetical Austrian decision maker must take a decision on the issue, based on an analysis of the environmental impact of the available options, i.e. landfill or incineration. The model reads a set of input data (e.g. waste inventories, emission factors for various compounds) and generates for each option a pressure-to-decision index PI. PI(I) quantifies how much the option (I) would impact on the environment. The model output is a function, Y, of the PIs for incineration and landfill, built in such a way as to suggest incineration for negative values of Y and landfill otherwise. Because most of the input factors to the analysis are uncertain, a Monte Carlo analysis is performed, propagating uncertainties to produce a distribution of values for Y (Figure 2.2). What makes this example instructive is that one of the 'factors' influencing the analysis is the way the model Y is built. This is not an uncommon occurrence in model based assessments. In this case, 'Finnish' and 'Eurostat' in Figure 2.2 refer to two possible choices of systems of indicators. The model also includes other

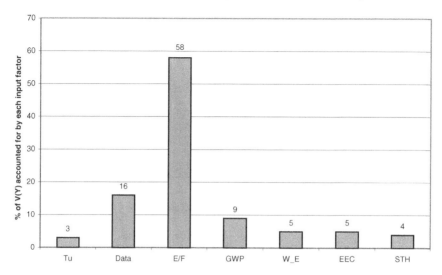

Figure 2.3   Variance-based decomposition of the output variable Y. The input factors (and their cardinality) are: E/F(1): trigger factor used to select randomly, with a uniform discrete distribution, between the Finnish and the Eurostat sets of indicators; TU (1), Territorial Unit: trigger that selects between two spatial levels of aggregation for input data; DATA (176), made up of activity rates (120), plus emission factors (37) plus National emissions (19); GWP (1), weight for greenhouse indicator (in the Finnish set): three time-horizons are possible: W_E (11), Weights for Eurostat indicators; EEC (1), approach for Evaluating Environmental Concerns: Target values (Adriaanse 1993) or expert judgement (Puolamaa *et al.*, 1996) and STH (1) is a single factor that selects one class of Finnish stakeholders from a set of eight possible classes. The factor E/F triggering the choice of the system of indicators accounts for 58% of the variance of Y.

methodological alternatives on choice of time horizons for impact assessment of waste disposals, weights for the sub-indicators and so on. As Figure 2.2 suggests, the main choice as to what system of composite indicator to use ('Finnish' or 'Eurostat') almost completely drives the answer. This is shown in the histogram of Figure 2.3, where a variance decomposition analysis has been used, but can be seen even better in Figure 2.4, where a Smirnov test is performed. The analysis indicates that, at the present state of knowledge, the issue is non-decidable. Furthermore, resources should not be allocated to obtaining a better definition of the input data (e.g. emission factors or inventories) but to reach a consensus among the different groups of experts on an acceptable composite indicator of environmental impact for solid waste.

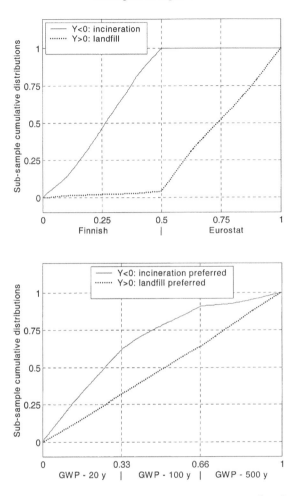

Figure 2.4 Smirnov test statistics (see Box 2.1 Smirnov) for the two cumulative sub-sample distributions obtained splitting the sample of each input factor according to the sign of the output Y. An unequivocal split between the solid and the dotted curves can be observed when the E/F input factor is plotted on the x-axis. When the GWP factor is plotted on the x-axis, the split between the two curves is less significant.

A model of fish population dynamics

A stage-based model was developed by Zaldívar *et al.* (1998) to improve understanding of the dynamics of fish ecosystems, and in particular the relative importance of environmental fluctuations and biological mechanisms in determining such dynamics. The model mimics data on scale deposition rates of small pelagic fish, i.e. sardine and anchovy, in different locations: the California current off western North America and the Benguela current off south-western Africa. Comparing geological data and simulation results, Zaldivar *et al.* showed that although environmental fluctuations can explain the magnitude of observed variations in geological recordings and catch data of pelagic fishes, they cannot explain the low observed frequencies. This implies that relevant non-linear biological mechanisms must be included when modelling fish population dynamics.

Despite the fact that the ecological structure of the model has been kept as simple as possible, in its final version the model contains over 100 biologically and physically uncertain input factors. With such a large number of factors, a sensitivity screening exercise may be useful to assess the relative importance of the various factors and the physical processes involved. The sensitivity method proposed by Morris (1991) and extended by Campolongo *et al.* (2003b) has been applied to the model. The model output of interest is the annual population growth. The experiment led to a number of conclusions as follows.

- The model parameters describing the population of sardines are not identified as substantially influential on the population growth. This leads to the question: 'Is this result reflecting a truly smaller role of this species with respect to the others or is it the model that should be revised because it does not properly reflect what occurs in nature?'

- A second conclusion is that the model parameters describing the inter-specific competition are not very relevant. This calls for a possible model simplification; there is no need to deal with a 100-factors model that includes inter-specific competitions when, to satisfy our objective, a lower level of complexity and a more viable model can be sufficient.

- Finally, a subset of factors that have the greatest effect on population growth (e.g. early life-stages parameters) is identified. This allows one to prioritise future research to invest time and effort in improving the estimates of these uncertain parameter values that control most of the output uncertainty (Factors Prioritisation Setting, see below).

The risk of a financial portfolio

Imagine that a bank owns a simple financial portfolio and wants to assess the risk incurred in holding it. The risk associated with the portfolio is defined as the difference between the value of the portfolio at maturity, when the portfolio is liquidated, and what it would have gained by investing the initial value of the portfolio at the market free rate, rather than putting it in the portfolio. The model used to evaluate the portfolio at each time is based on the assumption that the spot interest rate evolves on the market according to the Hull and White one-factor stochastic model (see details in Chapter 3).

Sensitivity analysis is performed on the financial portfolio model following a Monte Carlo (MC) filtering approach. The aim of MC filtering is to perform multiple model evaluations and then split the output values into two subsets of risk levels: those regarded as 'acceptable' (i.e. risk below a given threshold), and those considered 'unacceptable'. Consequently the input factors may be classified as 'behavioural' or 'not behavioural', depending on whether they lead to acceptable or unacceptable outputs. The test (see Box 2.1 Smirnov) is applied to each input factor to test whether the distributions of the 'behavioural' and 'not behavioural' values can be regarded as significantly different. The higher the Smirnov test value for an input factor, the higher is its influence on the model response. Although the MC filtering/Smirnov approach has some limitations, the main one being that it only captures first-order effects and cannot detect interactions among factors, its use in this setting provides some advantages. The idea of defining an 'acceptable' behaviour for the model is particularly apt for problems where the output is required to stay between certain bounds or below a given threshold. Furthermore, the method provides not only a measure of the importance of the input factors, but it also offers some indications of the type of relationship that link the input and output values. Our general advice is to use this approach in conjunction with quantitative global sensitivity analysis, as was done in the case of the financial portfolio, in order to obtain a complete picture of the problem. The Smirnov test and the variance-based techniques have been applied to the financial portfolio model to identify the input factors mostly responsible for possible losses.

The analysis outcome was also an assessment of how much risk results from uncertain input factors that are uncontrollable, such changes in interest rate, and how much is, in contrast, reducible because of factors that can be adjusted. The MC filtering analysis has provided an added value to the analysis by investigating the link between the portfolio performance and the number of times its composition is revised.

Box 2.1   SMIRNOV.

The Smirnov test is applicable when a qualitative definition for the 'good' or 'acceptable' behaviour of a model can be defined, for example, through a set of constraints: thresholds, ceilings or time bounds based on available information on the system. The steps for the analysis are as follows.

- Define a range for k input factors $X_i, i = 1, 2, \ldots k$, reflecting uncertainties in the model and make a number of Monte Carlo simulations. Each Monte Carlo simulation is associated with a vector of values of the input factors.

- Classify model outputs according to the specification of the 'acceptable' model behaviour [qualify a simulation as behaviour (B) if the model output lies within the constraints, non-behaviour $(\bar{B})$ otherwise].

- Classifying simulations as either B or $\bar{B}$, a set of binary elements is defined to allow one to distinguish between two sub-sets for each X_i: $(X_i|B)$ of m elements and $(X_i|\bar{B})$ of n elements [where $n + m = N$, the total number of Monte Carlo runs performed].

The Smirnov two-sample test (two-sided version) is performed *independently* for each factor.

Under the null hypothesis that the two probability density functions $f_m(X_i|B)$ and $f_n(X_i|\bar{B})$ are identical:

$$H_0 : f_m(X_i|B) = f_n(X_i|\bar{B})$$
$$H_1 : f_m(X_i|B) \neq f_n(X_i|\bar{B})$$

the Smirnov test statistic is defined by

$$d_{m,n}(X_i) = \sup_y || F_m(X_i|B) - F_n(X_i|\bar{B}) ||.$$

where F are marginal cumulative probability functions.

The question answered is: 'At what significance level α does the computed value of $d_{m,n}$ determine the rejection of H_0?'

A low level of α implies high values for $d_{m,n}$, suggesting that X_i is a key factor in producing the defined behaviour for the model.

A high level of α supports H_0, implying an unimportant factor: any value in the predefined range is equally likely to fall in B or $\bar{B}$.

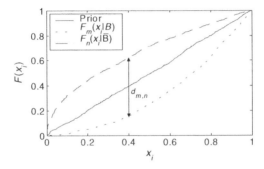

The Smirnov test will be more extensively discussed and applied in Chapter 6.

Two spheres

Imagine a model with six factors. Imagine that we want to estimate the model factors, using observations from the real system that the model attempts to describe. This would be formalised into the optimisation of a loss or a likelihood function, which, in the simplest case, will be a function of the mean square error between model simulations and observations. This setting is similar to the financial portfolio case, in which we identify and analyse the set of acceptable input factor configurations but, instead of filtering model runs according to the acceptability criteria, we 'rank' such runs according to a likelihood function, without discarding any of them. Let us also assume that the six factors interact in the model in such a way that the maximum likelihood (our optimum) lies on the surfaces of two three-dimensional spheres. The likelihood function to be analysed would therefore have the following form (in a log-scale):

$$f(X_1, \ldots, X_6) = -\left(\sqrt{X_1^2 + X_2^2 + X_3^2} - R_1\right)^2 \Big/ A_1$$
$$-\left(\sqrt{X_4^2 + X_5^2 + X_6^2} - R_2\right)^2 \Big/ A_2 \quad (2.1)$$

We pretend not to know the easy geometrical properties of the optimum, e.g. as if we only had a computational version of the model and the likelihood function came from a numerical routine comparing model runs and observations, such that we consider the optimisation in six dimensions instead of the more direct two radial dimensions. This is an example of an over-parameterised model, characterised by ill-defined parameters. This usually occur when trying to estimate mechanistic models with a very general functional form, usually in terms of sets of partial differential equations, defined according to first principles. Such models are usually more complex than the observed reality would allow one to define. This is reflected in a lack of identifiability of model parameters (factors); i.e. the optimum is usually not given by a 'maximum' in the model parameter space, but by a complex interaction structure in which many different combinations of the parameters are equally able to provide best fitting model simulations.

In spite of the apparent simplicity of the function (2.11), there is no simple technique that can identify the optimal factor's structure that maximises the likelihood function, consisting of two groups of three factors in a spherical symmetry. In fact, tools such as correlation analysis, principal component analysis and Monte Carlo filtering are ineffective to highlight the three-dimensional structures. Global sensitivity analysis insted allows the identification of the interaction structure caused by over-parameteterisation, i.e. the two disjoint groups of three factors each, even though SA cannot identify the spherical configuration. In fact, global SA allows the identification of the elements and groups characterising the interaction structure, but not the topological configuration of that structure. This result is very useful as it leads the analyst to a more efficient search in the two relevant subsets, which might allow them ultimately to elucidate the spherical geometry. This example shows the use of global SA in optimisation problems, in which global SA results are a useful ingredient in the search for complex optimum structures.

A chemical experiment

In this example we apply sensitivity analysis tools to model estimation/calibration problems, in conjunction with Bayesian techniques. A very simple chemical system is considered, consisting of the observation of the time evolution of an isothermal first-order irreversible reaction in a batch system $A \rightarrow B$ (Ratto *et al.*, 2001). We want to fit a kinetic model using a set of observations. We would like to know not only the optimum, but also the structure of the model parameters that allows a good fit. In this case we start from our prior beliefs on the model parameters and the acceptability is classified according to a loss or likelihood function, which, as in the two-spheres case, will be based on the residuals between model predictions and observations. As in the two-spheres example, we use global SA tools to identify the main properties of the acceptable set of model parameters (i.e. the optimal parameter structure). This also allows the identifiability of the parameters to be assessed, i.e. which parameter can be determined given the data. This permits a reduction of the dimension of the estimation problems, by ignoring/fixing the subset of factors classified as irrelevant by the sensitivity analysis. The two-spheres case study was designed to represent an over-parameterised model with a complex structure, in which the underlying interaction between factors is not elementarily detectable. In this case we show that even a very simple estimation problem can present aspects of over-parameterisation and interaction. So, again in this example we show how global SA can be useful as a first ingeredient in complex estimation/ calibration problems.

2.2 What is sensitivity analysis?

The term sensitivity analysis is variously interpreted in different technical communities, and problem settings. Thus a more precise definition of the terms demands that the output of the analysis be specified.

Until quite recently, sensitivity analysis was conceived and often defined as a local measure of the effect of a given input on a given output. This is customarily obtained by computing via a direct or indirect approach, system derivatives such as $S_j = \partial Y / \partial X_j$, where Y is the output of interest and X_j an input factor (Rabitz, 1989; Turanyi, 1990).

The local approach is certainly valuable for a class of problems that can be loosely defined as 'inverse', i.e. the determination of some physical parameters embedded into a complex model from experimental determination of observables that are further downstream in the model. A typical example is the determination of kinetic constants or quantum mechanic potentials from the yield rate of a chemical process (Rabitz, 1989). This approach is still quite widespread. As discussed in Saltelli (1999) most of the sensitivity analyses that can be found in physical science journals are local, sometimes inappropriately so.

The problem setting is different for practitioners involved in the analysis of risk (industrial, financial, etc.), decision support, environmental appraisal, regulatory compliance analysis, extended impact assessment, etc. For these the degree of variation of the input factors is material, as one of the outputs being sought from the analysis is a quantitative assessment of the uncertainty around some best estimate value for Y (uncertainty analysis). This can be achieved in simple cases by analytic expression or Taylor expansion, but is achieved most often and generally by Monte Carlo methods in conjunction with a variety of sampling strategies (see Helton, 1993). In this context, sensitivity analysis is aimed, amongst others, at priority setting, to determine what factor most needs better determination, and to identify the weak links of the assessment chain (those that propagate most variance in the output). Sensitivity analysis in this context is often performed using regression techniques, such as standardised regression coefficients. The analysis is performed by feeding the regression algorithm (such as ordinary least squares) with model input and output values. The regression algorithm returns a regression meta-model, whereby the output Y is described in terms of a linear combination of the input factors. The regression coefficient for a given factor plays the role of a sensitivity measure for that factor (Box 2.2 Regression in use).

Box 2.2   REGRESSION IN USE.

We start from our usual model of k factors $Y = f(X_1, X_2, \ldots, X_k)$, and imagine drawing a Monte Carlo sample from the input:

$$
\mathbf{M} = \begin{matrix}
x_1^{(1)} & x_2^{(1)} & \cdots & x_k^{(1)} \\
x_2^{(2)} & x_2^{(2)} & \cdots & x_k^{(2)} \\
\cdots & & & \\
x_1^{(N)} & x_k^{(N)} & \cdots & x_k^{(N)}
\end{matrix}
$$

Each row in $\mathbf{M}$ is a realisation from the multivariate joint probability distribution of the set $\mathbf{X}$, while a generic column j in $\mathbf{M}$ is a sample from the marginal distribution of the corresponding input factor X_j. We can compute the model $Y = f(X_1, X_2, \ldots, X_k)$ for each row in $\mathbf{M}$, obtaining a vector of model estimate.

$$
\mathbf{y} = \begin{matrix}
y^{(1)} \\
y^{(2)} \\
\cdots \\
y^{(N)}
\end{matrix}
$$

Let $\bar{x}_j$ and s_j be the mean and standard deviation of the sample of X_j in $\mathbf{M}$, and $\bar{y}$ and s_y the corresponding quantities in $\mathbf{y}$. We can now reckon in terms of standardised variables $\tilde{x}_j^{(i)} = (x_j^{(i)} - \bar{x}_j)/s_j$, $\tilde{y}^{(k)} = (y^{(i)} - \bar{y})/s_y$ and seek a regression model $(\hat{y}^{(i)} - \bar{y})/s_y = \sum_{j=1}^{k} \beta_j \tilde{x}_j^{(i)}$, where the β_js are determined by ordinary least square to minimise the sum of the squared $\varepsilon^{(i)}$, $\varepsilon^{(i)} = \hat{y}^{(i)} - y^{(i)}$.

The β_js are called standardised regression coefficients (SRC). These can be computed on most standard statistical packages and can be used as a simple and intuitive measure of the sensitivity of Y with respect to its input factors X_j. The effectiveness of the regression coefficients is conditional on the R_y^2 of the fit, defined as $R_y^2 = \sum_{i=1}^{N}(\hat{y}^{(i)} - \bar{y})^2 / \sum_{i=1}^{N}(y^{(i)} - \bar{y})^2$. If the fit of the regression is good, e.g. R_y^2 is larger than, say, 0.7, this means that the regression model is able to represent a large part of the variation of Y. This also means that the model is relatively linear. In such cases, the regression model is effective and we can base our sensitivity analysis on it. This is done taking the β_js, with their sign, as a measure of the effect of X_j on Y. We can order the factors by importance depending on the absolute value of the β_js.

One advantage of these methods is that in principle they explore the entire interval of definition of each factor. Another is that each 'effect' for a factor is in fact an average over the possible values of the other factors. Moreover, SRCs also give the sign of the effect of an input factor on the ouput, providing a simplified model of the input–output mapping. We call methods of this type 'global', to distinguish them from 'local' methods, where only one point of the factors' space is explored, and factors are changed one at a time. A disadvantage of regression based methods is that their performance is poor for non-linear models, and can be totally misleading for non-monotonic models. These methods are not discussed in this primer other than in Chapter 1.

In recent years global quantitative sensitivity analysis techniques have received considerable attention in the literature (special issues on the subject are RESS (1997, 2002), JSCS (1997) and CPC (1999). Many techniques have been developed that can be considered as a 'model free' extension of the regression methods above, as they can be applied even to non-linear, non-monotonic models.

Other global approaches are the First Order Reliability Methods, FORM, (Saltelli *et al.*, 2000a, p. 155) which are not described in this primer or the use of techniques derived from experimental design (see Saltelli *et al.*, 2000a, p. 51). This primer will describe just one selected screening technique, the method of Morris (Morris, 1991; Saltelli *et al.*, 2000a, p. 65), which is the most widely used, and it will then focus on model-free, variance based quantitative techniques as well as on Monte Carlo filtering approaches. As discussed later in this volume, the Morris method and the variance based measures are related to one another, while MC filtering can be used in conjunction with variance based methods.

The definition of sensitivity analysis that matches the content of this primer better is 'The study of how the uncertainty in the output of a model (numerical or otherwise) can be apportioned to different sources of uncertainty in the model input'. The rationale for our choice of a restricted subset of the universe of available sensitivity analysis methods is our belief that the selected techniques are best suited to meet a wide spectrum of applications and settings. Our motivation is in terms of 'desired properties', and 'possible settings', which are described next. A general schematic description of the steps to be followed to perform sensitivity analysis on a model, independently of the method being used, is given in Box 2.3, Steps for SA.

Box 2.3   STEPS FOR SA.

1. Establish what is the goal of your analysis and consequently define the form of the output function that answers your question(s) (remember that this should be a top level statement rather than the model output as it is).

2. Decide which input factors you want to include in your analysis. At this level, trigger parameters can be defined, allowing one to sample across model structures, hypotheses, etc. Moreover, in the case of multi-dimensional maps of factors, define the characterising parameters to represent them in the SA.

3. Choose a distribution function for each of the input factors. This can be:

(i) taken from the literature; or

(ii) derived from data by fitting an empirical distribution function; or

(iii) based on an expert's opinions;

(iv) chosen to be a truncated normal distribution, where truncation serves to avoid sampling outliers.

(v) In the case of triggers among, for example, different models, a 'Russian roulette' system with equal weights can be the first choice, but different weights can be given to different values of the trigger, when one has some prior information about a model structure/hypothesis being more likely than another.

(vi) Define a correlation structure between input factors, if appropriate.

4. Choose a sensitivity analysis method on the basis of the following.

(i) The questions that you are trying to address. For instance, you may face a screening problem or, in contrast, need a method, that is quantitative in order to be able to answer your final question.

(ii) The number of model evaluations that you can afford, on the basis of the model execution time. If, for instance, the number of input factors included in the analysis is high and the model is time-consuming, you are forced to choose a method that requires a low number of model evaluations, such as that proposed by Morris, or to group factors.

(iii) The presence of a correlation structure between input factors. When a screening problem is at hand, correlation should be dropped, to avoid useless complexity in the analysis, and introduced only for the subsequent quantitative analysis for the few important factors remaining.

5. Generate the input sample. This has the form of N strings of input factor values on which you evaluate the model. The sample is generated according to the method chosen for the sensitivity analysis. When using SIMLAB the sample generation follows the choice of method, the specification of the method's internal parameters, and the selected sample size.

6. Evaluate your model on the generated sample and produce the output, which contains N output values in the form specified in (1).

7. Analyse the model outputs and draw your conclusions, possibly starting a new iteration of the analysis.

2.3 Properties of an ideal sensitivity analysis method

We plan to use methods that are global and model-free, in the sense of being independent from assumptions about the model, such as linearity, additivity and so on. These methods must be capable of testing the robustness and relevance of a model-based analysis in the presence of uncertainties. Whenever possible, we would also like our methods to be quantitative. Our main choice is to work on variance-based methods, also known as importance measures, sensitivity indices or correlation ratios. We have used these in our worked example in Section 2.1. The Morris method, in its standard use, i.e. at low sample size, can be considered as qualitative. However, the variance based measures are quantitative, in principle, as long as the sample size is large enough and one can tell by how much factor a is more important than factor b. The choice of a qualitative versus quantitative method is driven by cost. The Morris method is much cheaper, in terms of model evaluations, than the variance based measures, though there are similarities in their interpretation (Chapter 4).

The desirable properties of sensitivity analysis, are as follows.

1. The ability to cope with the influence of scale and shape. The influence of the input should incorporate the effect of the range

Table 2.1   Properties of sensitivity measures.

	Property 1 Scale and shape	Property 2 Multi-dimensional averaging	Property 2 Model independence	Property 2 Grouping of factors
$S_j = \dfrac{\partial Y}{\partial X_j}$ (Local method, not applied in this primer)	N	N	N	Y
$S_i = SRC(Y, X_i)$ (Regression method, Box 2.2 Regression in use, not applied in this primer)	Y	Y	N	N
Morris (Chapter 4)	N/Y[a]	Y	Y	Y[b]
Variance based methods (Chapter 5)	Y	Y	Y	Y
Monte Carlo filtering (Chapter 6)	Y	Y	Y	N

[a] A coarse stratified sampling considering few levels in the quantile scale is possible with Morris' method, thus implying some coarse analysis of the influence of the scale and shape. Normally very few quantiles are used in Morris, see Chapter 4.
[b] See Campolongo *et al.* (2003).

of input variation and the form of its probability density function (pdf). It matters whether the pdf of an input factor is uniform or normal, and what the distribution parameters are.

2. To include multidimensional averaging. In a local approach to SA (e.g. $S_j = \partial Y / \partial X_j$), one computes the partial derivatives, as discussed above. This is the effect of the variation of a factor when all others are kept constant at the central (nominal) value. A global method should instead evaluate the effect of a factor while all others are also varying.

3. Being model independent. The method should work regardless of the additivity or linearity of the model. A global sensitivity measure must be able to appreciate the so-called interaction effect, which is especially important for non-linear, non-additive models. These arise when the effect of changing two factors is different from the sum of their individual effects as discussed in Chapter 1.

4. Being able to treat grouped factors as if they were single factors. This property of synthesis is essential for the agility of the interpretation of the results. One would not want to be confronted with an SA made of dense tables of sensitivity measures (see, for example, the decision analysis example at the beginning of this chapter).

Table 2.1 summarises the behaviour of various sensitivity measures with respect to properties (1–4).

Beside the properties above, we would like the setting for the SA itself to be as stringent as possible. It may well happen that using different measures of sensitivity, different experts obtain different relative ranking of the influence of the various input factors (see OECD, 1993 for an example). This happens if the objective of the analysis is left unspecified. Just as there are several definitions of risk (Risk Newsletter, 1987), there may be several definitions of importance. Below, we shall offer some alternative rigorous settings for SA that will help us in our analysis (Saltelli and Tarantola, 2002).

2.4 Defensible settings for sensitivity analysis

Uncertainty and sensitivity analyses are more often mentioned than practised. Anecdotal evidence puts the blame on model developers (i.e. ourselves occasionally). The modellers' overconfident attitudes might result in an under-estimation of predictive uncertainty and an incomplete understanding of the input–output relationship. A similar conceptual error in experimental sciences has been documented by Henrion and Fischhoff (1986). In computational sciences the issue might be at least as acute.

Another factor that may limit the application of sensitivity analysis is that it can be performed in many different ways. If one may obtain different orderings of the factors importance using different methods, why bother doing it? Importance is not per se a mathematical concept. Our answer to this question is that 'importance' must be defined at the stage of framing the analysis, as we discuss next.

Usually factors in a model follow a very asymmetric distribution of importance, few factors accounting for most of the output uncertainty with most factors playing little or no role. When this is the case, different methods may converge to the same result. On the other hand, a rigorous definition of importance is necessary, as the ordering of factors by importance may be an issue of great significance when the model is used, for example, in risk analysis or decision making.

In order to discuss this, we again assume a mathematical or computational model $Y = f(X_1, X_2, \ldots, X_k)$, where some of the input factors X_i are uncertain. Throughout this book k will be the number of factors whose variation is of interest. We know something about the range of uncertainty of these factors. This knowledge might come from a variety of sources: measurements, expert opinion, physical bounds or an analogy with factors for similar species/compounds. We may additionally have information (e.g., via observation) on the joint probability distribution of the factors.

The model may be used in a prognostic (forecasting) or diagnostic (estimating, calibrating) mode. In the former, all our knowledge about model input is already coded in the joint probability distribution of the input factors. In the latter, the input information constitutes a 'prior', and the analysis might be aimed at updating either the distribution of the input factors or the model formulation based on the evidence (see Chapter 6).

A 'forecast' mode of use for the model is assumed in the following unless otherwise specified. We select one among the many outputs produced by the given model and call this our output of interest. This might also be in the form of an averaged mean over more model outputs. The output of interest should be in the form of a single quantity, possibly a scalar Y, whose value is taken as the top-most information that the model is supposed to provide. This could be, for instance, the ratio of the value of an environmental pressure variable over the selected target value. It could be the maximum or averaged number of health effects in a given area and time span. It could be the estimated failure probability for a system in a given time span and so on. We express this by saying that a sensitivity analysis should not focus on the model output as such, but rather on the answer that the model is supposed to provide or on the thesis that it is supposed to prove or disprove. In $Y = f(X_1, X_2, \ldots, X_k)$, one does not need to assume f to be constant, as it is customary to propagate uncertainty through different model structures or formulations. In this case some of the input factors are triggers that drive the selection of one structure versus another, and f stands for the computational code where all this takes place.

Some of the factors can be the constituent parameters of an error model that has been built to characterise the uncertainty in (multi-dimensional) input data maps. The alternative of defining one input factor for each pixel in an input map would be impracticable (and useless) as we would have, say, one million input factors per input map. An example is given in Crosetto and Tarantola (2001). It is not impossible for a factor to be a trigger that drives the choice of one input data set versus another, where each set represents internally consistent but mutually exclusive parametrisations of the system. An example is given in Saltelli (2002), in which a trigger factor is set to select between alternative geological scenarios, each characterised and represented by a specific input file.

Let us assume that we are able to compute the model output as much as we like, possibly sampling from the best joint probability distribution of input that we can come up with. This procedure is called by some a parametric bootstrap, in the sense that we sample with replacement the factors that enter into a model and re-evaluate the model each time. Let us further assume, for simplicity, that each factor indeed has a true, albeit unknown, value. We know that often factors are themselves lumped entities called in as surrogates for some more complex underlying process, but we now assume that they are simply scalar variables imprecisely known because of lack of sufficient observations.

This clearly does not apply to stochastic uncertainties, such as the time of occurrence of an earthquake in a given area, although one might have frequency information for the area based on geological or historical records. Even in this case it is useful to think of the stochastic factor as possessing a true value, for the sake of assessing its importance relative to all other factors. We can at this point introduce our first setting for SA.

Factors Prioritisation (FP) Setting

The objective of SA is to identify the most important factor. This is defined as the one that, if determined (i.e., fixed to its true, albeit unknown, value), would lead to the greatest reduction in the variance of the output Y. Likewise, one can define the second most important factor and so on till all the factors are ranked in order of importance.

One might notice that we have made the concept of importance more precise, linking it to a reduction of the variance of the target function. It should also be noted that, in general, one would not be able to meet the objective of Setting FP, as this would imply that we know what the true value of a factor is. The purpose of Setting FP is to allow a rational choice under uncertainty.

Another thing worth noting about Setting FP, which will be elaborated below, is that it assumes that factors are fixed one at a time. This will prevent the detection of interactions, i.e., in adopting Setting FP, we accept the risk of remaining ignorant about an important feature of the model that is the object of the SA: the presence of interactions in the model. This point will be discussed further in Chapter 5.

The ideal use for the Setting FP is for the prioritisation of research; this is one of the most common uses of SA. Under the hypothesis that all uncertain factors are susceptible to determination, at the same cost per factor, Setting FP allows the identification of the factor that is most deserving of better experimental measurement in order to reduce the target output uncertainty the most. In order not to leave this setting just hanging here, we would like to say that Setting FP can be tackled using conditional variances such as $V(Y|X_i = x_i^*)$. As discussed in Chapter 1, this formula reads as: 'the variance of Y that is obtained when one factor, X_i, is fixed to a particular value, x_i^* '. The variance is taken over all factors that are not X_i, i.e. one might rewrite the formula as $V_{\mathbf{X}_{-i}}(Y|X_i = x_i^*)$, where $\mathbf{X}_{-i}$ indicates the vector of all factors but X_i. Because we do not normally know where to fix X_i, we go on to take the average of $V_{\mathbf{X}_{-i}}(Y|X_i = x_i^*)$ over all possible values of X_i, to obtain $E_{X_i}(V_{\mathbf{X}_{-i}}(Y|X_i))$, or $E(V(Y|X_i))$ in a more compact notation. Although $V(Y|X_i = x_i^*)$ might be either smaller or larger than $V(Y)$, depending on the values selected for x_i^*, $E(V(Y|X_i))$ is always smaller than $V(Y)$ (see Box 2.4 Conditional and unconditional variances). These measures and their estimation are described in Chapter 5.

Box 2.4   CONDITIONAL AND UNCONDITIONAL VARIANCES.

Let us consider the following model $Y = X_1 X_2^2$, where $X_1 \sim U(-0.5, 0.5)$ and $X_2 \sim U(0.5, 1.5)$. Scatter plots from a Monte Carlo simulation are shown below.

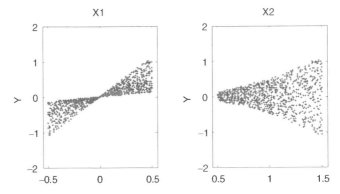

The analytic unconditional variance of Y is $V(Y) = 121/960 = 0.126$, whereas the conditional variances, shown in the figure below, are:

$$V(Y|X_1) = (61/180) X_1^2$$
$$V(Y|X_2) = (1/12) X_2^4.$$

It can be seen that the variance of Y conditional on X_2 is larger than $V(Y)$ for a large part of the support of X_2. Nonetheless, averaging over the conditioning argument, one obtains:

$$E[V(Y|X_1)] = 61/2160 = 0.0282 < V(Y) \Rightarrow X_1 \text{ influent}$$
$$E[V(Y|X_2)] = 121/960 = 0.126 = V(Y) \Rightarrow X_2 \text{ non-influent}$$

confirming that the inequality $E[V(Y|X_i)] \leq V(Y)$ always holds true. This is an example of the criticisms of Krykacz-Hausmann (2001) mentioned in Section 2.5. According to the first-order partial variance, the parameter X_2 is totally unimportant ($S_2 = 0$), because $E[V(Y|X_2)] = V(Y)$, whereas it is clear that by varying X_2 the variance of Y changes significantly. In such a case, the measure based on entropy would give a non-zero sensitivity index. This does not mean that variance based measures should be ruled out, because in this example it is clear that a practitioner would recover the effect of X_2 at the second order. Specifically, the pure interaction term would be: $V_{12} = V[E(Y|X_1, X_2)] - V[E(Y|X_1)] - V[E(Y|X_2)] = 0.0282$, i.e. 22.4% of $V(Y)$.

Factors Fixing (FF) Setting

This is concerned with the fixing of non-influential factors (see Sobol', 1990). The objective of this setting, which could also be labelled 'screening', is to identify the factor or the subset of input factors that we can fix at any given value over their range of uncertainty without significantly reducing the output variance. If such a factor or subset of factors are identified, the remaining ones, being varied within their own range, explain most of the unconditional variance.

This has implications in the process of simplifying complex models. The factors that do not influence the output can be fixed at their nominal values without any significant loss of information in the model. If one has prior beliefs about the importance of input factors, this setting can be used to prove or disprove a given model representation. The Setting FF can be treated using both the extended Morris and the variance-based techniques. We have mentioned already in Chapter 1 that a necessary and sufficient condition for factor X_i to be totally non-influential is $E(V(Y|\mathbf{X}_{-i})) = 0$. Application examples are given in Chapters 4, 5 and 6.

Variance Cutting (VC) Setting

This is a setting that we have found useful when SA is part of a risk assessment study. The objective of the analysis is the reduction of the variance of the output Y from its unconditional value $V(Y)$ to a lower pre-established threshold value. One must obtain a variance of Y that is equal to or smaller than a given target variance $V_r < V(Y)$ by simultaneously fixing the smallest number of factors. Even in this case we have to make an informed choice without knowing where the true values of the factors lie.

Also for Setting VC we are allowed only to make an informed choice, rather than finding the optimum for which the true factors' value would need to be known. Setting VC allows the factors to be fixed by groups, and the solution in this case can be influenced by the interactions between factors, if these are present. Even this setting can be treated using conditional variances, including conditional variances of higher order (conditioned on more than one factor, Chapter 5).

Factors Mapping (FM) Setting

In this setting, the realisations of the MC simulation of our output variable Y are cathegorised into two groups: for example, all those above a given percentile of $p(Y)$ and all those below, where $p(Y)$ is the empirical (MC-generated) distribution of the realisations. Alternatively the realisations of Y can be classified as either acceptable or non-acceptable by comparing them with either evidence or opinion. As discussed in Chapter 6, this is the setting of MC filtering, and the question addressed in sensitivity analysis for the FM setting is 'which factor is most responsible for producing realisations of Y in the region of interest?'. The FM setting takes its name from the backward mapping from output to input that is realised after Y is classified. Another possible question for Setting

FM is 'which factor is most responsible for splitting the realisations of Y into "acceptable" and "unacceptable"?' Setting FM is tackled among other methods by using the Smirnov test on the filtered versus unfiltered distributions of the factors (see Box 2.1 Smirnov and Chapter 6).

We shall mainly use Settings FP, FF, VC and FM in the present primer, although it is clear that many others are possible. Ultimately, the setting should be decided by the problem owner(s), and it may well be that the setting itself is debatable (i.e. do I care about the variance of Y or about Y upper 5th percentile? Is there a threshold for Y? How do I rank the importance of factors with respect to this?). Settings may well be audited, especially in a context where the analysis must meet participatory requirements.

On the other hand, we try to make the point that a setting must be defined for the analysis to be unambiguously implemented.

2.5   Caveats

In the previous section, we have assumed that one is interested in describing the output uncertainty in terms of its variance. In some decision contexts, there may be other measures that are more important, depending on the preferences of the owner of the problem. We may be concerned about shifts in central tendency (mean) of a model output attributable to an input factor, regardless of its contribution to the variance in the model. In OECD (1993) an analysis was performed by shifting the entire distribution of each input factor by a given (5%) fraction, and the resulting shift in the model output mean was used to rank the factors. This approach has some drawbacks, as discussed in Saltelli and Tarantola (2002). It is insensitive to model non-monotonicity and dependent on the fraction shift in the input distributions.

Krykacz-Hausmann (2001) has criticised the use of variance as a measure of output uncertainty, and suggested using entropy, H, instead, defined as either $H(Y) = -\int f(y) \ln(f(y)) \, dy$ or $H(Y) = -\sum p_i \ln(p_i)$ depending on whether the distribution of Y is continuous (f) or discrete (p). Krykacz-Hausmann's argument is that the largest uncertainty for Y should be that associated with a uniform distribution for Y in its range. With some intuitive examples (see for example Box 2.4 Conditional and unconditional variance), he argues that H is better than V in capturing this aspect of the problem. While the conditional variance $V(Y|X_i = x_i^*)$ can be larger than $V(Y)$, this does not occur with entropy, i.e.: $H(Y|X_i = x_i^*) < H(Y), \quad \forall X_i$.

As practitioners, we have struggled with possible alternatives to the variance. These have seemed to us to be associated with specific problems and are less convincing as a general method for framing a sensitivity analysis. As an example, some feel that it should be the output itself, for example Y, that is partitioned according to the influence of the different input factors. This is feasible (Sobol', 1990), and has been attempted (Sacks *et al.*, 1989). The function $Y = f(\mathbf{X})$ is partitioned, to a first approximation, into functions of just one factor, i.e. $Y \approx f_0 + \sum_i f_i(X_i)$. The analysis of sensitivity is then done by inspecting plots of each f_i versus its own variable X_i. The most important factor is the one whose f_i fluctuates the most, i.e. deviates the most from the mean value of Y (see Box 2.5 HDMR (High Dimensional Model Representations)). Even if we are not saying it, we are still judging upon contribution to the variance of Y.

Box 2.5   HDMR (HIGH DIMENSIONAL MODEL REPRESENTATION).

If one is interested in the output itself, for example Y, to be partitioned according to the influence of the different input factors, the function $Y = f(\mathbf{X})$ is partitioned, to a first approximation, into functions of just one factor, i.e. $Y \approx f_0 + \sum_i f_i(X_i)$. Terms of higher order may also be of interest.

For example, let us consider the Ishigami function (Ishigami and Homma, 1990):

$$Y = \sin X_1 + A\sin^2 X_2 + B X_3^4 \sin X_1, \quad \text{where } X_i \sim U(-\pi, \pi) \tag{1}$$

In this case, a possible closed decomposition is given by the following terms:

$$Y = f_0 + f_1(X_1) + f_2(X_2) + f_3(X_3) + f_{13}(X_1, X_3) \qquad (2)$$

where

$$
\begin{aligned}
f_0 &= A/2 = E(Y) \\
f_1(X_1) &= \sin X_1 \cdot (1 + B\pi^4/5) = E(Y|X_1) - f_0 \\
f_2(X_2) &= A\sin^2 X_2 - A/2 = E(Y|X_2) - f_0 \\
f_3(X_3) &= 0 = E(Y|X_3) - f_0 \\
f_{13}(X_1, X_3) &= B \sin X_1 \cdot (X_3^4 - \pi^4/5) \\
&= E(Y|X_1, X_3) - f_1(X_1) - f_3(X_3) - f_0.
\end{aligned}
\qquad (3)
$$

The first term of the decomposition is just the unconditional mean, whereas the remaining terms describe the deviation around the mean due to the various factors alone or through interaction. The last term of (3) represents a pure interaction term of the factors X_1 and X_3. This is the only non-zero interaction term of the Ishigami function, therefore the remaining ones do not appear in the decomposition (2).

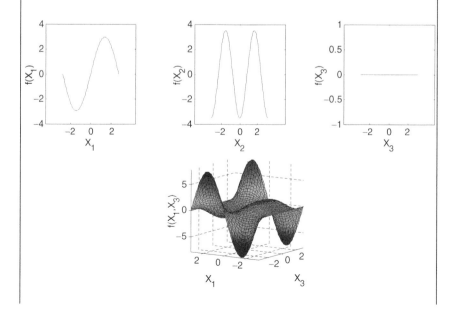

Plotting the various terms of the decomposition, one can visualise the contribution of each term to the model output Y. This is an example of what is called High Dimensional Model Representation (HDMR) (Rabitz *et al.*, 1999; Saltelli *et al.*, 2000a, pp. 199–223).

An alternative often suggested by practitioners is that of seeking to partition the mean, rather than the variance of Y, according to the input factors. However, a mean decomposition is suggested by users addressing questions such as (i) 'which factors determine the model output to be what it is, or (ii) 'at what values should I fix the input factors to obtain a target output value', rather than 'which factors cause the model output to vary the most'. While the latter is a typical SA question, the first two are more optimisation questions. Furthermore how could we practically construct a sensitivity measure based on a mean decomposition? Imagine that X_i is fixed to the value x_i^*. If one averages this, i.e. takes the mean value over X_i of $E(Y|X_i = x_i^*)$, then the unconditional mean $E(Y)$ is obtained ... which is not very inspiring. If one takes the variance over X_i of $E(Y|X_i = x_i^*)$, then one falls back on the sensitivity measure described in Chapter 5. In an optimisation framework we suggest using SA first to determine the subset of input factors driving most of the variation in model output, in order to reduce the problem dimensionality, and then to carry out a search on those factors to establish their optimal values.

In this framework, Monte Carlo filtering (Chapter 6) can be also effectively applied, specifically for question (ii) above. This is particularly relevant in such cases when, in dealing with complex models that are defined starting from first principles, a large set of ill-defined parameters is encountered when trying to estimate (optimise) parameters. For such ill-defined inverse problems, filtering techniques can be very useful to explore the parameter space pertaining to the complex (often multiple) optimum.

In conclusion, the choice of the method of sensitivity analysis is driven by the question that the owner of the problem is trying to answer. If one is capable of formulating a meaningful 'sensitivity' question, then one can perhaps find the appropriate tool to answer it, taking inspiration from the examples given in this primer.

We would like to end this section with a warning about the fallacies most often encountered in the literature, or from the authors' experience, on the subject of sensitivity analysis.

One is the improper use of local sensitivity analysis methods. A local sensitivity measure is one that looks at the relation between input and output at a specified point in the space of the inputs, such as a simple $(\partial Y / \partial X_i)$ or normalised $(\partial Y / \partial X_i \bar{x}_i / \bar{y})$ derivative. As mentioned in Chapter 1, these have a wide spectrum of applications, such as solving inverse problems or to accelerate, by Taylor expansion and approximation, the computation of complex models in the neighbourhood of a set of initial or boundary conditions (see Grievank, 2000). Although one frequently sees in the literature differential analysis used to assess the relative importance of factors in the presence of finite ranges of variations for the factors, this is a bad practice. On the contrary, the Morris method suggested in Chapter 4 looks at incremental ratios such as $(\Delta Y / \Delta X_i)^{(\mathbf{x}_j)}$, but these are taken at different points, $\mathbf{X}_j$s, in the space of the input factors. The mean and the standard deviation of the $(\Delta Y / \Delta X_i)^{(\mathbf{x}_j)}$ over the points $\mathbf{X}_j$ that are explored are used to detect the influential factors.

Similarly there might be an improper use of regression methods. Regressing model output on model input using a linear regression algorithm, such as, for example, ordinary least squares, produce regression coefficients for the factors (such as standardised regression coefficients) that can be used as coefficients of sensitivity. Yet this is only useful if the regression is successful, i.e. if the model coefficient of determination R_y^2 is high (e.g. 0.7 or more). When instead R_y^2 is low (e.g 0.3 or less), we say that the regression analysis is used inappropriately. One may still have a use for the regression analysis but one should not use it for sensitivity analysis.

We also noticed at times the tendency of some modellers to devise ad hoc methods for the sensitivity analysis of their model. While this may sometime be justified by the application and by the setting as discussed above, one should be careful that the analysis answers the question that is relevant to the problem, and that it is not designed to confirm the modeller's judgement upon the relative importance of factors.

The analysis should be as much as possible parsimonious and synthetic in order to be meaningful. If helps if the analysis answers a question relative to the use made of the model, rather than to the model itself.

The preparation of the input sample for a case where the input factors are not independent of each other can be laborious (Chapter 5). When the factors are independent, important simplifications in the computational procedures can be realised. Ignoring these would result in an analysis that is unnecessarily expensive.

In our experience, sensitivity analysis is fairly complex and iterative in practice. It most often uncovers errors in the model and it is rarely the case that the outcome does not contain at least one unexpected result. Its contribution to the quality of the model is evident and should justify the effort needed to implement it. In the European Union, key policy directives must undergo a so-called Extended Impact Assessment (EIA), a cross-sectorial analysis of the costs and benefits of the various options (including the do-nothing option) that specifically includes sensitivity analysis (EC, 2002). Similar concerns are evident in the US regulations, for example in the Environmental Protection Agency White Paper on model acceptability (EPA, 1999).

3 TEST CASES

In this chapter we illustrate a number of practical problems where the application of a sensitivity analysis exercise has provided considerable added value to the study. The following sections describe the test cases and highlight the conclusions that could be drawn thanks to the analysis. The aim is also to stress which sensitivity approach is the most appropriate for a given problem. The methods used are merely mentioned but not described.

3.1 The jumping man. Applying variance-based methods

We like extreme and exciting sports. This is the exercise for us: bungee jumping. We are standing on a platform; far below is the asphalt, at a distance H that we cannot quantify well (probably 40 m, most likely 50 m, perhaps 60 m).

The challenge is to choose the best bungee cord for our mass, i.e. the one that allows us to almost touch the ground below, thus giving us a real thrill. Other less suitable cords will either be fatal for us or will give us a very short ride (not exciting at all!). Therefore, which cord should we use? There are many types of cords comprising different numbers of strands (say, from 10 to 50). More strands mean a stronger, but less exciting, cord. We consult with our friends and decide that a good range for this variable is (20, 40). The other variable driving the oscillation is our mass. Unfortunately, we do not remember our mass exactly; it has been two months since we last weighted ourselves and then we were about 70 kg. At first glance we might well be between 67 kg and 74 kg now. Perhaps we

Sensitivity Analysis in Practice: A Guide to Assessing Scientific Models A. Saltelli, S. Tarantola, F. Campolongo and M. Ratto © 2004 John Wiley & Sons, Ltd. ISBN 0-470-87093-1

don't have a strong desire to risk our life testing the system life. We need a 'virtual' bungee jump, i.e. a simple simulator of our system.

Nonetheless, we still have a poor knowledge of the variables of the system, and we may want to test what effects such uncertainty can have on the outcome of our jump.

What are the variables of interest to quantify the risk of failure? Qualitatively, we would like to know what is the probability of success for our jumps (SJ). SJ might be 100% (i.e. no risk at all) but the jump will not be at all exciting. Therefore, we want to identify an indicator for both risk and excitement: this can be the minimum distance to the asphalt during the oscillation ($h_{\min}$). This is a typical optimisation problem, where SJ has to be maximised and $h_{\min}$ minimised (with the constraint that $h_{\min} > 0$).

We are interested in:

1. estimating the empirical distribution of $h_{\min}$ and SJ for all the combinations of the input factors' values;

2. selecting for which input factor we would gain a better level of accuracy in order to have the highest reduction of the uncertainty of the risk. The information attained here will become a priori information for us to use in the real world.

We can employ global uncertainty and sensitivity analysis to answer both these questions. The model is represented by the simple function

$$h_{\min} = H - \frac{2Mg}{k_{\mathrm{el}}\sigma}, \tag{3.1}$$

which is the solution of the linear oscillator equation. H is the distance of the platform to the asphalt [m], M is our mass [kg], σ is the number of strands in the cord, g is the acceleration of gravity [m/s^2], k_{el} is the elastic constant of one strand [N/m], which is considered fixed here at 1.5.

On the basis of the thinking above we assume the following.

- The uncertainty on H can be suitably represented by a uniform distribution with a minimum value of 40 m and maximum value of 60 m.

- The uncertainty on M can be represented by a uniform distribution with a lower bound of 67 kg and an upper bound of 74 kg.

- The uncertainty on σ can be represented by a uniform distribution with a lower bound of 20 strands and an upper bound of 40 strands.

We run an uncertainty analysis based on a random sample of 1000 points. Figure 2.1 displays the empirical histogram of h_{min}, and also enables us to estimate SJ. In fact, in 974 cases out of 1000 the jump is successful.

Our knowledge of the analytical formulation of the model (i.e. Equation (3.1)) indicates that:

- the model is linear on factor H,
- the model is linear on the ratio M/σ,
- but not on M and σ separately.

For this reason, we adopt a model-free method of sensitivity analysis. We want to identify the variables that most influence the model output in terms of setting FP (Chapter 2). Hence, we estimate the first-order sensitivity indices using a method based on the decomposition of the model output variance. The analysis shows that the first-order sensitivity indices for our variables are:

$$S_H = 0.44$$
$$S_M = 0.01$$
$$S_\sigma = 0.55$$

The reader can reproduce these results with SIMLAB. The analysis shows that the number of strands in the cord (σ) is the variable where we should direct effort in order to reduce the uncertainty on h_{min}. We also understand that the distance platform–asphalt (H) is important. Hence, we should try to get a better knowledge for this variable, for example, by doing an indirect estimation of the distance (e.g., by measuring the time taken by a small stone to touch the ground below). At the same time, we should not waste time in improving the accuracy of our weight, as its effect on the uncertainty of h_{min} is negligible.

While the knowledge of H can be improved as remarked above, the uncertainty on σ can be completely removed by selecting a given cord. Let us select $\sigma = 30$ strands and repeat the uncertainty analysis by generating 1000 random points over the space (H, M) and evaluating the model. We get a safe jump $(SJ = 100\%)$ with h_{min} ranging between 1.3 m and 24.7 m above the asphalt.

A final remark: if we add up the three sensitivity indices of the first test, we can appreciate the degree of additivity of the model. The sum of the three indices is exactly equal to 1. This means that, in spite of its analytic formulation, the model is almost fully additive, i.e. no interaction emerges between the variables M and σ. Readers are invited to try tuning the input distributions, for example, changing the width of the uniform densities, to identify when interactions emerge more clearly.

3.2 Handling the risk of a financial portfolio: the problem of hedging. Applying Monte Carlo filtering and variance-based methods

Imagine that a bank has issued a financial contract, namely a *caplet*, a particular type of European option whose value depends on the curve of the interest rate. The way interest rates evolve through time is unknown, and therefore by selling a caplet the bank is facing the risk associated with interest rate movements.

Assume that the bank wants to offset such a risk. The goal is not to make a profit but to avoiding the risk exposure of having issued the option. In finance this is called the problem of *hedging*.

The bank buys a certain amount of FRAs (Forward Rate Agreements) that are contracts that, by behaving in an opposite way to that of the caplet with respect to changes in interest rates, are capable of offsetting the caplet risk exposure. The amount of FRAs purchased is such that the overall bank's portfolio, made by the caplet and the FRAs, is insensitive (or almost insensitive) to interest rates movements. The portfolio is said to be *delta neutral*, delta indicating the type of risk being hedged (offset).

As time passes the portfolio tends to lose risk neutrality and again to become sensitive to interest rates changes. Maintaining risk neutrality would require the portfolio to be continuously revised, as the amount of FRAs needed to maintain neutrality changes with time as the interest rate curve moves. As in practice only a limited number of portfolio revisions are feasible and also because each revision implies a cost, a hedging error is generated and, at maturity, when the caplet expires and the portfolio is liquidated, the bank may incur a loss. The goal of the bank is to quantify the potential loss.

The hedging error is defined as the difference between the value of the portfolio at maturity and what would have been gained by investing the initial value of the portfolio at the interest rate prevailing on the market (the market free rate). Note that when the error is positive it means that, although failing to maintain risk neutrality, the bank is making a profit. In contrast, when the error is negative, the bank is losing money.

In order to compute the hedging error at maturity, we need to be able to evaluate the portfolio at any time, or, in other words, we need to be able to price the financial contracts included in the portfolio, the caplet and the forward rate agreements. To this end we make use of the Hull and White one-factor model (Rebonato, 1998, p. 281), which assumes the interest rates evolution through time to be driven by only one factor, the spot interest rate r_t, evolving as:

$$dr_t = \mu(t, r_t)\, dt + \sigma(t, r_t)\, dW_t \qquad (3.2)$$

with

$$\mu(t, r_t) = \Theta(t) - ar_t \quad \text{and} \quad \sigma(t, r_t) = \sigma. \qquad (3.3)$$

The terms $\mu(t, r_t)$ and $\sigma(t, r_t)$ are respectively the drift and the standard deviation (volatility) of the spot rate, $\{W_t : t > 0\}$ is a (standard) Wiener process, a is the constant mean-reverting parameter, and the choice of the time dependent function $\Theta(t)$, e.g. a polynomial, can be directly determined from the initial yield curve, i.e. from all the information available at $t = 0$.

The hedging error depends upon a number of factors such as, for instance, the number of portfolio revisions performed, or other parameters related to the assumptions made on the way that interest rates evolve with time. The number of revisions is a factor that is unknown and controllable, in the sense that the bank has the faculty to decide how many revisions to perform but a priori does not know what the optimal number is. The cost incurred to revise the portfolio partially offsets the benefit deriving from increasing the number of portfolio adjustments. The parameters related to the interest rate evolution are not only unknown, but are also uncontrollable.

Uncertainty analysis estimates the uncertainty in the hedging error taking into account the uncertainty affecting the input factors, both those controllable and uncontrollable. The estimated error assumes the form of a distribution of values, rather than being a unique value, and elementary statistics, such as the mean, standard deviations, and percentiles, are used to describe its features. Uncertainty analysis shows the average error, its range of variation, and, for instance, the 99th percentile of the distribution that can be interpreted as the maximum loss that the portfolio's owner faces with a probability of 99%. In financial literature this percentile is referred to as *value at risk*.

Once the bank has quantified the potential loss, the focus is on what determines this loss to be what it is. The goal of sensitivity analysis is to identify what is causing the loss and to obtain indications on how to reduce it.

Sensitivity analysis is performed on the financial portfolio model following a Monte Carlo filtering approach, and using the Smirnov test to assess the relative importance of the input factors. In a Monte Carlo filtering approach the range of a model output responses is categorized into two groups: one designated 'acceptable' behaviour, and the other 'unacceptable', where 'acceptable' is defined to suit the nature of the problem at hand. The Smirnov test is used to compare statistically the sets of input values that lead to acceptable behaviour and those that do not. Those factors for which the sample distribution functions are found to be significantly different in the two sub sets are identified as being the most important in determining the specified behaviour (in this case the error to be over or under a certain threshold). For the important factors, graphical analysis is also advisable: a histogram of the input sample distribution leading to acceptable output values may provide information on the type of relationship existing between

Table 3.1 Input factors distributions for the Monte Carlo filtering analysis.

Factor	Description	Distribution
a	Constant mean-reverting parameter in Equations (3.2–3.3)	Normal (0.1; 0.01)
σ	Volatility of the spot rate in Equations (3.2–3.3)	Uniform (0.005;0.055)
N. rev.	Number of portfolio revisions	Discrete uniform (0,5,8,17,35,71)
ε	Trigger selecting among ten possible paths of the spot rate, from which the yield curve is determined	Discrete uniform $(1, \ldots, 10)$

input and output values. Details on Monte Carlo filtering and on the Smirnov test can be found in Box 2.1 Smirnov (Chapter 2) and in Chapter 6.

To complete the analysis, several possible scenarios have been considered, each scenario corresponding to a different value of the transaction costs incurred when revising the portfolio composition.

Our analysis considers four factors: a factor representing the variability of the dynamics of the evolution of the interest rate through time (ε); the number of portfolio revisions to be performed (N. rev.); and the parameters a and σ of the Hull and White model of the spot rate. Their assumed statistical distributions are given in Table 3.1.

The type of analysis carried out on the financial test case allowed some conclusions to be drawn, both in terms of relative importance of the input factors in determining the potential loss incurred by the bank, and in terms of the strategy to adopt to reduce this loss. The conclusions drawn are as follows.

- If there are no transaction costs, the highest percentage of acceptable values is obtained when the maximum number of portfolio revisions are performed. As expected, when transaction costs are introduced, it is more appropriate to reduce the number of revisions. In each scenario, analysis of the distribution of the acceptable values as a function of the number of portfolio revisions performed provides an indication of the optimal number of revisions that the bank should carry out in order to reduce the potential loss.

- In each transaction costs scenario, the sensitivity analysis executed via the Smirnov test has indicated that the parameter a in the model describing the interest rate evolution is irrelevant when compared with the others. This may lead to the conclusion that, when calibrating the model on the market prices to recover the 'best values' for a and σ, the effort to recover a may be unnecessary. However, as underlined in Chapter 6, the MC filtering/Smirnov approach's main limitation is that it only captures first-order effects and cannot detect interactions between factors. A variance-based analysis (see Chapter 5) has underlined that, although less important than the aleatory uncertainty due to the unknown interest rates dynamics, ε, or less important than the number of portfolio revisions performed (N. rev.), the model parameter a has a non-negligible total effect, mostly due to its interaction with other factors. Therefore its value cannot be fixed.

Although the test case shown here is very simple and takes into account only a limited number of uncertain input factors, it is sufficient to prove that uncertainty and sensitivity analyses are valuable tools in financial risk assessment. Uncertainty analysis quantifies the potential loss incurred by the bank and, in particular, the maximum potential loss, a variable that is often of interest in this context. Sensitivity analysis identifies the relative importance of the sources of the incurred risk. In particular, it splits the risk into the amount, which is not reducible, that is due to the intrinsic uncertainty that the financial analyst cannot control (e.g. that associated with the interest rate changes), and the amount that in principle may be reduced by making proper choices for 'controllable' input factor values (such as the number of portfolio revisions to carry out).

The example has also shown that, in the case of financial risk analysis, the Monte Carlo filtering/Smirnov approach represents an attractive methodology. The definition of 'acceptable' model behaviour is in fact particularly indicated when addressing risk problems where the output is required to stay below a given threshold. Furthermore, this analysis addresses not only the relative importance of the sources of uncertainty in the analysis outcome but also the type of relationship that links the input and the output values, which is the main question addressed by financial analysts.

Nevertheless, we recommend the use of this approach in conjunction with variance-based techniques, as these may overtake the limits of the MC/Smirnov analysis.

3.3   A model of fish population dynamics. Applying the method of Morris

This section describes a test case taken from Saltelli *et al.* (2000a, p. 367). A sensitivity analysis experiment is applied to a model of fish population developed to improve the understanding of the dynamics of fish ecosystems, and in particular the relative importance of environmental fluctuations and biological mechanisms in determining such dynamics.

Zaldivar *et al.* (1998) addressed the problems of modelling the dynamics of fish ecosystems by using Lotka–Volterra non-linear differential equations and stage-based discrete models. A number of models were developed to mimic data on scale deposition rates of small pelagic fish, i.e. sardines and anchovies, in different locations: the California current off western North America and the Benguela current off south-western Africa. Comparing geological data and simulation results, Zaldivar *et al.* (1998) showed that although environmental fluctuations can explain the magnitude of observed variations in geological recordings and catch data of pelagic fishes, they cannot explain the low observed frequencies. This implies that relevant non-linear biological mechanisms must be included when modelling fish population dynamics.

The class of model chosen as the most apt to describe fish population dynamics, is that of stage-based models (Caswell, 1989). Stage-based models, in contrast to continuous ordinary differential equations that ignore population structure and treat all individuals

as identical, integrate population dynamics and population structure very clearly. These models are really useful when the life cycle
is described in terms of size classes or development stages rather
than age classes. In a stage-based model, it is assumed that vital
rates depend on body size and that growth is sufficiently plas-
tic that individuals of the same age may differ appreciably in
size.

The basis of a stage-based model is the matrix $\mathbf{A}$ describing the
transformation of a population from time t to time $t+1$:

$$\mathbf{n}_{t+1} = \mathbf{A}\,\mathbf{n}_t \tag{3.4}$$

where A has the following structure:

$$\mathbf{A} = \begin{bmatrix}
P_1 & m_2 & m_3 & \cdots & & m_q \\
G_1 & P_2 & 0 & \cdots & & 0 \\
0 & G_2 & P_3 & 0\cdots & & 0 \\
\cdots & & & & \cdots & \\
0 & 0 & 0\cdots & & G_{q-1} & P_q
\end{bmatrix} \tag{3.5}$$

where $\mathbf{n}_t$ is a vector describing the population at each stage at time
t, P_i is the probability of surviving and staying in stage i, G_i is the
probability of surviving and growing into the next stage, and m_i
is the maternity per fish per unit time (days), $i = 1, 2, \ldots, q$.

Both P_i and G_i are functions of the survival probability, p_i, and
the growth probability, γ_i (Caswell, 1989):

$$\begin{aligned}
P_i &= p_i(1 - \gamma_i) \\
G_i &= p_i\gamma_i
\end{aligned} \tag{3.6}$$

where $p_i = e^{-z_i}$, $\gamma_i = (1 - p_i)p_i^{d_i-1}/1 - p_i^{d_i}$, z_i is the daily instan-
taneous mortality rate (IMR) and d_i is the duration (days) within
the ith stage.

The model developed by Zaldivar et $al.$ (1998) involves three
species: sardines (I), anchovies (J), and mackerel (K), and 13 different life stages, larvae (1–4), juvenile (5–9), and adult (10–13). Interspecies competition at the larval and juvenile stage is also taken
into account. The model therefore has the form of a block matrix
whose three central blocks contain the three population matrices
$\mathbf{A}_1, \mathbf{A}_2, \mathbf{A}_3$ respectively for the three species (sardines, anchovies

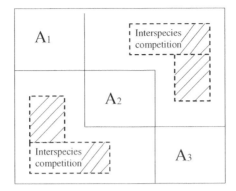

Figure 3.1   Schematic layout of the stage-based matrix model.

and mackerels), and the rest contains the parameters relating to larvae and juvenile interspecies competition (see Figure 3.1).

The model also includes diffusion to a refuge, which is modelled by doubling the dimensions of the matrix to include the free systems plus a common term for allowing the movement from one patch to the other. The diffusion is only allowed to the adult population.

Despite the fact that the ecological structure of the model has been kept as simple as possible, in its final version the model contains over 100 biological and physical factors. Values for these parameters can be found in the literature (see Saltelli *et al.*, 2000a, p. 373). The values used in this book are specified in Tables 3.2–3.4. The best parameters were chosen to produce a dominant eigenvalue λ in the matrix **A**, which represents the population growth rate, equal to 1, i.e. the population is stationary.

In order to assess the relative importance of the various factors and physical processes involved, a sensitivity analysis is performed on the model. The output variable of interest is λ^{365}, the annual population growth rate. The large number of factors imposes a choice of method that is not too computationally expensive, such as the screening design proposed by Morris.

Results of the sensitivity experiment contributed to an improvement in our understanding of the fish population dynamics and the merits and limits of the model employed. Conclusions that could be drawn from the analysis include the following.

Table 3.2 Stage-specific life-history factor values of the Pacific Sardine (see Saltelli et al., 2000a, p. 373).

Stage	Size (mm)		Daily natural mortality			Duration (days)			Daily fecundity		
	Min.	Max.	Min.	Best	Max.	Min.	Best	Max.	Min.	Best	Max.
Egg	Hatch	4	0.31	0.72	2.12	1.4	2.5	3.9	0	0	0
Yolk-sac larvae	4	10	0.394	0.6698	0.971	1.4	3.1	3.9	0	0	0
Early larvae	10	35	0.1423	0.2417	0.3502	5	11	21	0	0	0
Late larvae	35	60	0.057	0.0964	0.139	20	35	50	0	0	0
Early juvenile	60	85	0.029	0.056	0.081	17	25	40	0	0	0
Juvenile I	85	110	0.0116	0.0197	0.0285	30	50	80	0	0	0
Juvenile II	110	135	0.0023	0.004	0.0058	80	110	146	0	0	0
Juvenile III	135	160	0.0016	0.0028	0.004	105	146	185	0	0	0
Juvenile IV	160	185	0.0012	0.0022	0.0032	110	170	220	0	0	0
Prerecruit	185	210	0.0006	0.0011	0.0015	110	175	220	0	80	161
Early adult	210	235	0.0006	0.0011	0.0015	190	381	570	286	389	489
Adult	235	260	0.0006	0.0011	0.0022	400	663	920	730	946	1114
Late adult			0.0006	0.0011	0.0022	1908	2773	3473	1064	1688	3123

Table 3.3 Stage-specific life-history factor values of the Northern anchovy (see Saltelli et al., 2000a, p. 373).

Stage	Size (mm)		Daily natural mortality			Duration (days)			Daily fecundity		
	Min.	Max.	Min.	Best	Max.	Min.	Best	Max.	Min.	Best	Max.
Egg			0.12	0.231	0.45	1.4	2.9	3.9	0	0	0
Yolk-sac larvae	hatch	4	0.19	0.366	0.59	1.4	3.6	3.9	0	0	0
Early larvae	4	10	0.187	0.286	0.345	8	12	23	0	0	0
Late larvae	10	35	0.047	0.0719	0.087	35	45	71	0	0	0
Early juvenile	35	60	0.0009	0.02796	0.017	45	62	100	0	0	0
Late Juvenile	60	85	0.0029	0.0044	0.0053	60	80	138	0	0	0
Prerecruit	85	110	0.002	0.0031	0.0037	200	287	632	0	10.5	19.4
Early adult	110	135	0.0011	0.0021	0.0036	750	1000	1250	143.8	199.2	230.7
Late adult	135	160	0.0011	0.0021	0.0036	1000	1250	1500	284.2	448.4	529.0

Table 3.4   Stage-specific life-history factor values of the chub mackerel (see Saltelli et al., 2000a, p. 374).

Stage	Size (mm)		Daily natural mortality			Duration (days)			Daily fecundity		
	Min.	Max.	Min.	Best	Max.	Min.	Best	Max.	Min.	Best	Max.
Egg			0.126	0.240	1.614	6.43	16	21.14	0	0	0
Early larvae	4	10	0.020	0.150	0.360	3.03	23	33.74	0	0	0
Late larvae	10	35	0.016	0.050	0.079	29.08	89	121.2	0	0	0
Early Juvenile	35	60	0.0009	0.020	0.055	20	64	144.5	0	0	0
Juvenile	60	110	0.0009	0.010	0.045	17	128	289	0	0	0
Late Juvenile	110	135	0.0016	0.027	0.045	45	64	144.5	0	0	0
Prerecruit	135	225	0.0005	0.00056	0.0018	190	346	570	288.2	411.7	452.9
Early adult	225	300	0.0005	0.00056	0.0018	400	663	920	691.7	988.14	1086.1
Late adult	300	340	0.0005	0.00056	0.0018	1908	2773	3473	1165.4	1664.9	1831.4

1. Parameters involved in interspecies competition are not very relevant. This may lead to a model simplification as, at least for our objective function, it seems that there is no need to include this part of the model.

2. Parameters of early life-stages have the greatest effect on population growth. This confirms our expectations, as it is natural to think that early life stages strongly affect the population growth, and can therefore be seen as a guarantee of the model quality.

3. Fecundity factors are not very significant for any species at any life stage. This is a good indication of how to prioritise future research. Money and effort should be devoted to the measurements and estimate of other parameters rather than fecundity factors.

4. Parameters related to sardines are not amongst the most important. This is also very valuable information, as it may call for a revision of the model structure. Is the population of sardine really less important than the other two species in determining the overall equilibrium of the three-species population or have we somehow failed in our modelling process and should make some revisions?

3.4 The Level E model. Radionuclide migration in the geosphere. Applying variance-based methods and Monte Carlo filtering

Level E was used both as a benchmark of Monte Carlo computation (Robinson and Hodgkinson, 1987; OECD, 1989) and as a benchmark for sensitivity analysis methods (OECD, 1993). This test case has been extensively used by several authors, see Saltelli and Tarantola (2002) for a review. The model predicts the radiological dose to humans over geological time scales due to the underground migration of radionuclides from a nuclear waste disposal site. The scenario considered in the model tracks the one-dimensional migration of four radionuclides (^{129}I and the chain $^{237}Np \rightarrow ^{233}U \rightarrow ^{229}Th$) through two geosphere layers characterised by different hydro-geological properties. The processes being considered in the model are radioactive decay, dispersion, advection and the chemical reaction between the migrating nuclide and the porous medium. The repository is represented as a point source. Some time after the steel canister containing the waste has lost its integrity (the time of containment failure is indicated by T), the release of radionuclides to the geosphere depends only on the leach rates ($k_{(.)}$) and the initial inventory ($C_{(.)}$). The source term for ^{129}I is given by:

$$\frac{\partial C_I}{\partial t} = -\lambda_I C_I, \qquad t \le T$$
$$\frac{\partial C_I}{\partial t} = -\lambda_I C_I - k_I C_I, \qquad t > T \tag{3.7}$$

where C_I (mol) is the amount of ^{129}I, and λ_I (yr^{-1}) and k_I(yr^{-1}) are the decay rate and the leaching rate for ^{129}I. The initial condition is $C_I(t = 0) = C_I^0$, that is, the amount of ^{129}I at the time of vault closure (see Table 3.5). The source term for ^{237}Np, the first element of the chain, is described by

$$\frac{\partial C_{Np}}{\partial t} = -\lambda_{Np} C_{Np}, \qquad t \le T$$
$$\frac{\partial C_{Np}}{\partial t} = -\lambda_{Np} C_{Np} - k_C C_{Np}, \qquad t > T \tag{3.8}$$

in which the parameter k_C represents the leaching rate for the radionuclides of the chain. The source term for ^{233}U is given by

$$\frac{\partial C_U}{\partial t} = -\lambda_U C_U + \lambda_{Np} C_{Np}, \qquad t \le T$$
$$\frac{\partial C_U}{\partial t} = -\lambda_U C_U + \lambda_{Np} C_{Np} - k_C C_U. \qquad t > T \tag{3.9}$$

The source term for ^{229}Th is similarly described by

$$\frac{\partial C_{Th}}{\partial t} = -\lambda_{Th} C_{Th} + \lambda_U C_U, \qquad t \le T$$
$$\frac{\partial C_{Th}}{\partial t} = -\lambda_{Th} C_{Th} + \lambda_U C_U - k_C C_{Th}. \qquad t > T \tag{3.10}$$

Table 3.5 List of input factors for the Level E model.

Notation	Definition	Distribution	Range	Units
T	Containment time	Uniform	$[100, 1000]$	yr
k_I	Leach rate for iodine	Log-Uniform	$[10^{-3}, 10^{-2}]$	mol/yr
k_C	Leach rate for Np chain nuclides	Log-Uniform	$[10^{-6}, 10^{-5}]$	mol/yr
$V^{(1)}$	Water velocity in the first geosphere layer	Log-Uniform	$[10^{-3}, 10^{-1}]$	m/yr
$l^{(1)}$	Length of the first geosphere layer	Uniform	$[100, 500]$	m
$R_I^{(1)}$	Retention factor for iodine in the first layer	Uniform	$[1, 5]$	—
$R_C^{(1)}$	Retention factor for the chain elements in the first layer	Uniform	$[3, 30]$	—
$v^{(2)}$	Water velocity in the second geosphere layer	Log-Uniform	$[10^{-2}, 10^{-1}]$	m/yr
$l^{(2)}$	Length of the second geosphere layer	Uniform	$[50, 200]$	m
$R_I^{(2)}$	Retention factor for iodine in the second layer	Uniform	$[1, 5]$	—
$R_C^{(2)}$	Retention factor for the chain elements in the second layer	Uniform	$[3, 30]$	—
W	Stream flow rate	Log-Uniform	$[10^5, 10^7]$	m³/yr
C_I^0	Initial inventory for ^{129}I	Constant	100	mol
C_{Np}^0	Initial inventory for ^{237}Np	Constant	1000	mol
C_U^0	Initial inventory for ^{233}U	Constant	100	mol
C_{Th}^0	Initial inventory for ^{229}Th	Constant	1000	mol
w	Water ingestion rate	Constant	0.73	m³/yr
β_I	Ingestion-dose factor for ^{129}I	Constant	56	Sv/mol
β_{Np}	Ingestion-dose factor for ^{237}Np	Constant	6.8×10^3	Sv/mol
β_U	Ingestion-dose factor for ^{233}U	Constant	5.9×10^3	Sv/mol
β_{Th}	Ingestion-dose factor for ^{229}Th	Constant	1.8×10^6	Sv/mol

The migration through the geosphere is the core of the model. The migration of ^{233}U is governed by:

$$R_U^{(k)} \frac{\partial F_U^{(k)}}{\partial t} = v^{(k)} d^{(k)} \frac{\partial^2 F_U^{(k)}}{\partial x^2} - v^{(k)} \frac{\partial F_U^{(k)}}{\partial x} - \lambda_U R_U^{(k)} F_U^{(k)} + \lambda_{Np} R_{Np}^{(k)} F_{Np}^{(k)} \tag{3.11}$$

where U stands for the isotope ^{233}U, Np stands for ^{237}Np, (k) refers to geosphere layer number k (1 or 2), R_i is the retardation coefficient for nuclide i (dimensionless), $F_i(x, t)$ is the flux (amount transported per unit time) of nuclide i in the geosphere at position x and time t (mol/yr), $v^{(k)}$ is the water travel velocity in the kth geosphere layer (m/yr), $d^{(k)}$ is the dispersion length in the kth geosphere layer (m), and λ_i is the decay constant of nuclide i (yr^{-1}).

A similar equation holds for ^{229}Th:

$$R_{Th}^{(k)}\frac{\partial F_{Th}^{(k)}}{\partial t} = v^{(k)}d^{(k)}\frac{\partial^2 F_{Th}^{(k)}}{\partial x^2} - v^{(k)}\frac{\partial F_{Th}^{(k)}}{\partial x} - \lambda_{Th}R_{Th}^{(k)}F_{Th}^{(k)} + \lambda_U R_U^{(k)}F_U^{(k)}.$$
(3.12)

To simplify the model structure, the retardation coefficients R_U, R_{Np}, R_{Th} were replaced, in the Level E exercise, by a single parameter R_C. The equation for ^{129}I is:

$$R_I^{(k)}\frac{\partial F_I^{(k)}}{\partial t} = v^{(k)}d^{(k)}\frac{\partial^2 F_I^{(k)}}{\partial x^2} - v^{(k)}\frac{\partial F_I^{(k)}}{\partial x} - \lambda_I R_I^{(k)}F_I^{(k)}. \quad (3.13)$$

A similar equation holds for ^{237}Np provided that the index I is replaced by Np. The modelling of the biosphere is extremely simplified: the dose to the most exposed individual of a hypothetical critical group is computed via an ingestion factor and the water consumption rate. The radiological dose (measured in Sv/yr) from nuclide i is given by

$$Dose_i(t) = \beta_i\frac{w}{W}F_i^{(2)}(l^{(2)}, t), \quad i = {}^{129}I, {}^{237}Np, {}^{233}U, {}^{229}Th$$
(3.14)

where β_i is an ingestion-dose conversion factor and is assumed fixed, $F_i^{(2)}(l^{(2)}, t)$ is the flux at the end of the second layer (the output to the biosphere), w denotes the drinking water requirement for an individual in the most exposed critical group, and W is the stream flow rate. The quantity of interest in this study is the annual radiological dose due to the four radionuclides

$$Y(t) = \sum_i Dose_i(t)$$
(3.15)

The simulated time frame for applications presented in this book ranges from 2×10^4 to 9×10^6 years. The predictive uncertainty about $Y(t)$ is due to uncertainties in model parameters, both intrinsic (i.e., stochastic), such as the time to canister failure, or due to our poor knowledge of the system (i.e., epistemic), such as a poorly known kinetic parameter. The twelve uncertain input factors are listed in Table 3.5 together with a set of constant parameters.

The probability distributions for the factors have been selected on the basis of expert judgement. Such data refer to the original formulation of Level E (OECD, 1989), to which we refer in our analyses in Chapter 5.

Although experts were aware of the existence of correlations, for example, between R_I and R_C for each layer of the geosphere, these were omitted in the original system specifications for the difficulty in handling the estimation of sensitivity measures for correlated input (at that time, the very same concept of sensitivity analysis was unclear). In other analyses of Chapter 5 we have acknowledged the correlation between inputs to explain the behaviour of the physical system better. A significant vault failure can induce high leaching for both iodine and the chain elements, and vice versa. Also, high release coefficients for iodine should be accompanied by high release coefficients for the chain elements, within a given geosphere layer. The geochemical properties in the two layers of the geosphere are likely to be correlated, as well as the water flows in the two layers. In addition, the time to containment failure is likely to be correlated to the flow in the first layer, since corrosion will be faster if the flow is faster. The correlation pattern used in Chapter 5 is given in Table 3.6. The set of correlation values was defined by consulting with the authors of the benchmark (Robinson, 2000).

Table 3.6 Configuration for correlated input of the Level E model.

Pairs of correlated factors	Rank Correlation
k_I, k_C	0.5
$R_I^{(1)}, R_C^{(1)}$	0.3
$R_I^{(2)}, R_C^{(2)}$	0.3
$T, v^{(1)}$	-0.7
$v^{(1)}, v^{(2)}$	0.5
$R_I^{(1)}, R_I^{(2)}$	0.5
$R_C^{(1)}, R_C^{(2)}$	0.5

The executable of a Fortran 77 code of the Level E model is available for use within SIMLAB. The code solves the system of equations above numerically (Crank Nicholson was used by the authors, see Prado *et al.* (1991) and Crank (1975)) with the constant parameters given at the bottom of Table 3.5. The system could also be tackled using an algorithm for accurate numerical inversion of the solution in the Laplace space to obtain the solution in real space (Robinson and Hodgkinson, 1987).

The results of the analysis are described in detail in Chapters 5 and 6. The dynamics of the system is strongly non-linear and the relationship between the input factors and the model output is also non-monotonic. The most critical time point is at $t = 2 \times 10^5$ yr, where non-linearities, non-monotonicities and interactions between model parameters dominate the model behaviour.

In terms of the FF setting, the sensitivity analysis shows that the parameters of the steel canister, i.e. the containment time (T) and the leach rates for both iodine and for the nuclides of the neptunium chain $(K_I$ and $K_C)$ are non-influential over all the time range. So, they can be fixed in a subsequent analysis and the dimensionality of the space of the input factors can be reduced from 12 to 9.

The largest values of first-order effects are obtained for the water speed in the first geosphere layer $(v^{(1)})$, the stream flow rate (W) and the length of the first geosphere layer $(l^{(1)})$. However, the sum of all the first-order indices is less than 0.25 across all the time range. The output variance is driven mostly by interactions between the factors.

When a correlation structure is considered for the input factors, the sensitivity pattern is rather different, although W and $v^{(1)}$ still explain the largest part of the variance in the output. In terms of setting FP, W and $v^{(1)}$ are the factors that deserve more consideration in terms of uncertainty reduction.

In Chapter 6 we present an application of the Monte Carlo filtering approach at the crucial time point $t = 2 \times 10^5$ yr, in order to identify which parameters mainly drive the occurrence of extremely high values of the radiological doses (FM setting). The result is that a combination of low values of $v^{(1)}$ and W, and high values of $l^{(1)}$ contribute to high doses.

3.5   Two spheres. Applying variance based methods in estimation/calibration problems

Let us consider the two-spheres example introduced in Chapter 2, a case of estimation of an over-parameterised model (Ratto, 2003). The estimation ultimately consists of the optimisation of a loss function or likelihood, which is usually a function of the mean squared error between model simulations and observations. We assumed that the model factors interact in such a way that the optimum lies on the surface of two three-dimensional spheres, such as:

$$\left(\sqrt{X_1^2 + X_2^2 + X_3^2} - R_1\right)^2 \Big/ A_1 + \left(\sqrt{X_4^2 + X_5^2 + X_6^2} - R_2\right)^2 \Big/ A_2 = 0.$$

$$(3.16)$$

Moreover, we also assume not to know the easy geometrical properties of the optimum, but we only have a computational version of the model and of the likelihood function to optimise.

This model is over-parameterised, since only two parameters would be identifiable (the two radii), but six parameters have to be optimised; as a result many different combinations of the parameters are equally able to provide best fitting model simulations. Such combinations lie on the two three-dimensional spheres.

Let us fix the coefficients:

$$R_1 = R_2 = 0.9$$
$$A_1 = A_2 = 0.001$$

and assume the six input factors having prior distributions $N(0, 0.35)$. Starting from the prior assumptions, we want to analyse and search the optimal structure of model parameters for the 'black-box' function

$$f(X_1, \ldots, X_6) = -\left(\sqrt{X_1^2 + X_2^2 + X_3^2} - R_1\right)^2 \Big/ A_1$$
$$-\left(\sqrt{X_4^2 + X_5^2 + X_6^2} - R_2\right)^2 \Big/ A_2, \quad (3.17)$$

which can be seen as the kernel of a log-likelihood function obtained from comparisons between model simulations and observations, which has to be maximised.

If we apply tools such as correlation analysis, principal component analysis and Monte Carlo filtering we have no chance of highlighting the three-dimensional structures.

On the other hand, if we apply global (variance based) sensitivity methods, we can identify the interaction structure. In fact, the sensitivity indices tell us that:

- only terms up to the third order are non-zero, implying that the largest interaction structure has dimension three;

- among the third-order terms, only the third-order terms of the groups $[X_1, X_2, X_3]$ and $[X_4, X_5, X_6]$ are non-zero, suggesting that the key interaction structure is given by two subsets of three factors;

- third-order closed sensitivity indices of the two disjoint groups $[X_1, X_2, X_3]$ and $[X_4, X_5, X_6]$ sum up exactly to 1, implying that the interaction structure is exactly decomposable into the two groups!

The only limit is that SA tools cannot identify the spherical configuration: global SA tools enable identification of the elements and the groups characterising the interaction structure, but not the topological configuration of that structure. This result guides the analyst to a more efficient search in the two relevant subsets, which might allow one to ultimately elucidate the spherical geometry, showing that global SA results are useful as a 'prior' to the search of complex optimum structures.

This example and the following one (a chemical experiment) can also be seen as a 'mix' of settings described in Chapter 2:

- the Factors Mapping (FM) Setting, in which categorisation between acceptable and unacceptable behaviour is not obtained through the filtering, but through the 'labelling' of each Monte Carlo run according to the loss function/likelihood value;

- the Factors' Prioritisation (FP) Setting, in which taking the variance decomposition of the loss/likelihood function, the modeller is addressed to the subset of factors driving the acceptable behaviour of the model;

- the Factors' Fixing (FF) Setting, in which, always taking the variance decomposition of the loss/likelihood function, the modeller can ignore a subset of irrelevant factors in the calibration procedure.

3.6 A chemical experiment. Applying variance based methods in estimation/calibration problems

Let us assume that we are performing a set of experiments in a laboratory with the aim of studying the time evolution of an isothermal first-order irreversible reaction in a batch system $A \rightarrow B$ (Ratto et al., 2001). We also assume that we want to fit a kinetic model using the set of observations. We would like to know not only the optimum, but also the structure of the model parameters that allow a good fit. The first-order chemical process is described by the following differential equation:

$$\frac{dy_B}{dt} = ky_A \qquad (3.18)$$

where

$$y_i = \frac{n_i}{n_T} = \frac{n_i}{n_A^0 + n_B^0} \qquad (3.19)$$

$i = A, B$, is the dimensionless concentration, and k is the chemical kinetics rate constant.

The solution to this ordinary differential equation leads to:

$$y_B(t) = 1 + \left(y_B^0 - 1\right)\exp(-kt) \qquad (3.20)$$

A pseudo-experiment has been simulated, by considering the following conditions for the chemical system:

$$
\begin{aligned}
k &= k_\infty \exp(-E/RT) \\
k_\infty &= 2.5e5 \ \text{s}^{-1} \\
E/R &= 5000 \ \text{K} \\
T &= 300 \ \text{K} \\
[k &= 0.014 \ \text{s}^{-1}] \\
y_B^0 &= 0.1
\end{aligned}
\qquad (3.21)
$$

Table 3.7 Prior distributions of the kinetic model.

Factor	Distribution
k_∞	U[0, 5e5]
E	U[4500, 5500]
y_B^0	U[0, 0.3]

where k_∞ is the Arrhenius pre-exponential factor, E is the activation energy, R is the gas constant, T is the absolute temperature and y_B^0 is the initial concentration of B in the reactor.

To simulate observations, a normally distributed error has been added to the analytical behaviour defined in Equation (3.20), with zero mean and standard deviation 0.05.

We consider a vector of three parameters to be estimated: $X = [k_\infty, E, y_B^0]$. We start from our prior beliefs on the model parameters, formalised in the prior distributions presented in Table 3.7.

The acceptability of model factors is then classified according to a loss, a likelihood or a weighting function, which, as in the two-spheres case, will be based on the errors between model predictions and observations. We assume one can measure the concentration of B and we define the weighting function as (see Chapter 6 for details):

$$f\left(x_1^{(i)}, x_2^{(i)}, x_3^{(i)}\right) = \left(\frac{1}{\sigma^{(i)2}}\right)^\alpha, \quad i = 1, \ldots N \qquad (3.22)$$

where $\sigma^{(i)2} = 1/2 \cdot N_{obs} \sum_{t=1}^{N_{obs}} (\hat{y}_{Bt}(X = x^{(i)}) - y_{Bt})^2$ is the mean square error, N is the number of Monte Carlo runs performed, N_{obs} is the number of observations available, $X = [X_1, \ldots, X_3]$ is the vector of input factors, $\hat{y}_{Bt}(X = x^{(i)})$ and y_{Bt} are the simulated and observed time evolutions respectively of B.

As in the two-spheres example, we use global SA tools to identify the main properties of the acceptable set of model parameters (i.e. the optimal parameter structure). This also allows one to assess the identifiability of the parameters by highlighting those more clearly driven by data. This allows the dimension of the estimation problems to be reduced by ignoring/fixing the subset of factors classified as irrelevant by the sensitivity analysis.

The two-spheres case study was designed to represent an over-parameterised model, with a complex structure, in which the underlying interaction between factors is not elementarily detectable. In this case we show that even a very simple estimation problem can present aspects of over-parameterisation and interaction. In fact, the strong interaction between k_∞ and E is a well-known feature in the estimation of chemical rate constants (see e.g. Bard, 1974).

Two types of output have been considered: the physical output $y_B(t)$ and the weighting function arising from comparisons between model runs and observations.

Physical output
The sum of the first-order indices of the three model factors is never less than 0.86 and the initial condition y_B^0 has a non-negligible effect only at the very beginning of the simulation. Therefore little interaction is revealed by the analysis of the physical output, which simply singles out the importance of *both* kinetic factors and the irrelevance of the initial condition for most of the simulation.

Weighting function
The first-order sensitivity indices are much smaller than the main effects for the physical output and their sum is less than 0.2, implying that none of the four parameters is clearly identifiable from data. By analysing the total effect indices, very high sensitivity is detected for the chemical kinetics factors, implying that the behavioural runs are driven by an *interaction* between them. On the other hand, the influence of the initial condition is also small in terms of total effect.

From global sensitivity analysis results, we could conclude that:

- the initial condition y_B^0 is unimportant and therefore unidentifiable from data: i.e. any value in the prior distribution is equally likely to allow a good fit to observations;

- the chemical rate factors mainly drive the model fit to the experimental data;

- on the other hand, the chemical rate factors cannot be precisely estimated, since the absolute values of the first-order indices are small, leaving the main contribution to the output variance to interaction terms;

- the model is over-parameterised since there is a large difference between main and total effects.

In this example we also show that, changing the data set, by adding measurements at different temperatures, the identifiability/ over-parameterisation issues can drastically change: this allows us to show that the same model may or may not be over-parameterised according to the evidence with which it is compared.

3.7 An analytical example. Applying the method of Morris

The analytical example presented here is taken by Morris (1991). The model contains twenty input factors and has the following form:

$$y = \beta_0 + \sum_{i=1}^{20} \beta_i w_i + \sum_{i<j}^{20} \beta_{i,j} w_i w_j + \sum_{i<j<l}^{20} \beta_{i,j,l} w_i w_j w_l$$

$$+ \sum_{i<j<l<s}^{20} \beta_{i,j,l,s} w_i w_j w_l w_s \tag{3.23}$$

where $w_i = 2 \times (x_i - 1/2)$ except for $i = 3, 5$, and 7, where $w_i = 2 \times (1.1 x_i/(x_i + 0.1) - 1/2)$. Coefficients with relatively large values are assigned as

$$\beta_i = +20 \quad i = 1, \ldots, 10; \qquad \beta_{i,j} = -15 \quad i, j = 1, \ldots, 6;$$

$$\beta_{i,j,l} = -10 \quad i, j, l = 1, \ldots, 5; \quad \beta_{i,j,l,s} = +5 \quad i, j, l, s = 1, \ldots, 4.$$

The remaining first-and second-order coefficients are independently generated from a normal distribution with zero mean and unit standard deviation; the remaining third-and fourth-order coefficients are set to zero.

    This simple analytical model has been used to test the performance of the Morris screening method and proves its capabilities to distinguish between factors that have negligible effects, linear and additive effects, or non-linear or interaction effects.

Results showed that, considering both the Morris measures μ and σ, one can conclude that:

(i) the first ten factors are important;

(ii) of these, the first seven have significant effects that involve either interactions or curvatures;

(iii) the other three are important mainly because of their first-order effect.

These results are in line with what is expected by looking at the analytical form of the model and at the factor values, thus confirming the effectiveness of the Morris method in determining the relative importance of the model input factors.

4 THE SCREENING EXERCISE

4.1 Introduction

Mathematical models are often very complex, computationally expensive to evaluate, and involve a large number of input factors. In these cases, one of the aims in modelling is to come up with a *short list* of important factors (this is sometimes called the principle of parsimony or Occam's razor). The question to address is: 'Which factors – among the many potentially important ones – are really important?'

Answering this question is important for a number of reasons. When a few important factors are identified, the modeller may choose to simplify the model structure by eliminating parts that appear to be irrelevant or he may decide to proceed with model lumping and extract a simpler model from the complex one. The identification of the input factors driving most of the variation in the output is also a mean of quality assurance. If the model shows strong dependencies on factors that are supposed not to be influential, or the other way around, one may rethink the model and eventually decide to revise its structure. Furthermore, additional studies may be devoted to improving the estimates of the most influential factors, so as to increase the accuracy of model predictions.

To identify the most important factors from among a large number, the choice of a well-designed experiment is essential. The experiment must be designed to be *computationally cheap*, i.e.

Sensitivity Analysis in Practice: A Guide to Assessing Scientific Models A. Saltelli, S. Tarantola, F. Campolongo and M. Ratto © 2004 John Wiley & Sons, Ltd. ISBN 0-470-87093-1

requiring a relatively small number of model evaluations. Screening designs fulfil this requirement. These designs are conceived to deal with models containing tens or hundreds of input factors efficiently. As a drawback, these methods tend to provide qualitative sensitivity measures, i.e. they rank the input factors in order of importance, but do not quantify how much a given factor is more important than another.

Screening designs are a convenient choice when the problem setting is that defined in Chapter 2 as Factors' Fixing (FF). In the FF setting the objective is to identify the subset of input factors that can be fixed at any given value over their range of uncertainty without significantly reducing the output variance. The screening methods provide a list of factors ranked in order of decreasing importance, allowing the modeller to identify the subset of less influential ones.

Screening techniques have been applied to several practical simulation studies in different domains, providing good results. In general, screening designs perform better when the number of important factors in the model is small compared with the total number of factors. In other words, they perform better under the assumption that the influence of factors in the model is distributed as the wealth in nations, i.e. it follows Pareto's law, with a few, very influential factors and a majority of non-influential ones. In practice this is often verified and the results of screening exercises are generally rather satisfactory.

Several screening designs have been proposed in the literature (for a review see Saltelli *et al.* (2000b, p. 65)). In this chapter we shall focus on the design proposed by Morris (1991), and on some extensions of it (Campolongo *et al.*, 2003), as we believe this design to be the most appealing in several problem settings.

The method of Morris varies one-factor-at-a-time and is therefore referred to as an OAT method. Each input factor may assume a discrete number of values, called *levels*, which are chosen within the factor range of variation. Two sensitivity measures are proposed by Morris for each factor: a measure μ that estimates the overall effect of the factor on the output, and a measure σ that, according to Morris, estimates the ensemble of the second- and higher-order effects in which the factor is involved (including curvatures and interaction effects). The Morris measure, μ, is obtained by computing a number, r, of incremental ratios at different points $\mathbf{x}^{(1)}, \ldots, \mathbf{x}^{(r)}$ of the input space, and than taking their average. The number, r, of selected points is called the *sample size* of the experiment. Here we also describe a third measure, μ^*, proposed by Campolongo *et al.* (2003), which is a revised version of the Morris μ. μ^* is very successful in ranking factors in order of importance and performs capably when the setting is that of Factor's Fixing.

The method illustrated in this chapter is simple, easy to implement, and the results are easily interpreted. It is economic in the sense that it requires a number of model evaluations that are linear in the number of model factors. As a drawback, the method relies on a sensitivity measure, called the *elementary effect*, which uses incremental ratios and is apparently a local measure. However, the final measure, μ and μ^*, are obtained respectively by averaging several elementary effects and their absolute values computed at different points of the input space, so as to lose the dependence on the specific points at which the elementary effects are computed. In this sense, as it attempts to explore several regions of the input space, the method can be regarded as global.

Other screening methods that it is worth mentioning are: the design of Cotter (1979), the Iterated Fractional Factorial Designs, IFFDs (Andres and Hajas, 1993), and the sequential bifurcation proposed by Bettonvil (Bettonvil, 1990; Bettonvil and Kleijnen, 1997). However, with respect to each of these methods, the Morris's design has the benefit of a greater applicability. While the design of Cotter performs well when factors do not have effects that cancel each other out, the IFFD is recommended when only a restricted number of factors is important, and sequential bifurcation is ideal when factor effects have *known signs* (which means that the analyst knows whether a specific individual factor has a positive or negative effect on the simulation response); the Morris's design does not rely on restricted assumptions and is therefore model independent.

A description of these methods and a discussion on their properties can be found in Saltelli *et al.* (2000b, p. 65).

4.2 The method of Morris

The guiding philosophy of the Morris method (Morris, 1991) is to determine which factors may be considered to have effects which are (a) negligible, (b) linear and additive, or (c) non-linear or involved in interactions with other factors. The experimental plan proposed by Morris is composed of individually randomised 'one-factor-at-a-time' experiments; the impact of changing one factor at a time is evaluated in turn.

In order to illustrate this experimental plan, assume that the k-dimensional vector $\mathbf{X}$ of the model input has components X_i each of which can assume integer values in the set $\{0, 1/(p-1), 2/(p-1),\ldots,1\}$. The region of experimentation, Ω, will then be a k-dimensional p-level grid.[1]

The method suggested by Morris is based on what is called an *elementary effect*. The elementary effect for the ith input is defined as follows. Let Δ be a predetermined multiple of $1/(p-1)$. For a given value $\mathbf{x}$ of $\mathbf{X}$, the elementary effect of the ith input factor is defined as

$$d_i(\mathbf{x}) = \frac{[y(x_1,\ldots,x_{i-1},x_i+\Delta,x_{i+1},\ldots,x_k) - y(\mathbf{x})]}{\Delta} \quad (4.1)$$

where $\mathbf{x} = (x_1, x_2, \ldots, x_k)$ is any selected value in Ω such that the transformed point $(\mathbf{x} + \mathbf{e}_i\Delta)$, where $\mathbf{e}_i$ is a vector of zeros but with a unit as its ith component, is still in Ω for each index $i = 1,\ldots,k$.

The finite distribution of elementary effects associated with the ith input factor, is obtained by randomly sampling different $\mathbf{x}$ from Ω, and is denoted by F_i. The number of elements of each F_i is $p^{k-1}[p - \Delta(p-1)]$. Assume for instance that $k=2$, $p=5$, and $\Delta = 1/4$, for a total number of 20 elements for each F_i. The five-level greed in the input space is represented in Figure 4.1. The total number of elementary effects can be counted from the grid by simply keeping in mind that each elementary effect relative to a factor i is computed by using two points whose relative distance in the coordinate X_i is Δ.

---

[1] In practical applications, the values sampled in Ω are subsequently rescaled to generate the actual values assumed by the input factors.

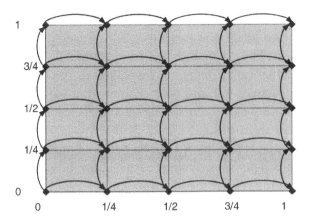

Figure 4.1 Representation of the five-level grid ($p = 5$) in the two-dimensional input space ($k = 2$). The value of Δ is $1/4$. Each arrow identifies the couple of points needed to compute one elementary effect. The horizontal arrows identify the 20 elementary effects relative to X_1, while the vertical ones identify the 20 elementary effects relative to X_2.

Campolongo *et al.* (2003) proposed that the distribution of the absolute values of the elementary effects, namely G_i, should also be considered. The examination of the distributions F_i and G_i provides useful information about the influence of the ith input factor on the output.

Here we take as the most informative sensitivity measures μ^*, the mean of the distribution G_i, and σ, the standard deviation of F_i. μ^* is used to detect input factors with an important overall influence on the output. σ is used to detect factors involved in interaction with other factors or whose effect is non-linear.

Note that in the original work of Morris (Morris, 1991) the two sensitivity measures proposed were respectively the mean, μ, and the standard deviation, σ, of F_i. However, choosing Morris has the drawback that, if the distribution, F_i, contains negative elements, which occurs when the model is non-monotonic, when computing the mean some effects may cancel each other out. Thus, the measure μ on its own is not reliable for ranking factors in order of importance. It is necessary to consider at the same time the values of μ and σ, as a factor with elementary effects of different signs (that cancel each other out) would have a low value of μ but a considerable value of σ that avoids underestimating the factors' importance. For interpreting results by simultaneously taking into account the two sensitivity measures, Morris suggested a graphical representation. The estimated mean and standard deviation of each sample of elementary effects are displayed in the (σ, μ) plane (see examples in Figures 4.4 and 4.5). The plotted values may thus be examined relative to each other to see which input factor appears to be the most important.

When the goal is to rank factors in order of importance by making use of a single sensitivity measure, our advice is to use μ^*, which by making use of the absolute value, avoids the occurrence of effects of opposite signs.

The mean of the distribution F_i, which comes out at no extra computational cost, can still be used to detect additional information on the signs of the effects that the factor has on the output. If the mean of F_i is high, it implies not only that the factor has a large effect on the output but also that the sign of this effect is always the same. If, in contrast, the mean of F_i is low, while the mean of G_i is high, it means that the factor examined has effects of different signs depending on the point in space at which the effect is computed.

To examine the effects due to interactions we use the original measure proposed by Morris and consider the standard deviation of the distribution F_i. An intuitive interpretation of its meaning is the following. Assume that, for factor X_i, we get a high value of σ. This means that the elementary effects relative to this factor are significantly different from each other, i.e. the value of an elementary effect is strongly affected by the choice of the point in the input space at which it is computed, i.e. by the choice of the other factor's values. In contrast, a low σ indicates very similar values of the elementary effects, implying that the effect of X_i is almost independent of the values taken by the other factors.

The sensitivity measures preferred here are therefore μ^*, the mean of the distribution G_i, and σ, the standard deviation of F_i. If we attempt to make a comparison with the variance-based measures proposed in Chapter 5, we see that μ^* is the best parallel of the total sensitivity index S_{T_i}. In fact, if we were to express μ^* in terms of variance operators, we would write $\mu^* = E[\psi\,(Y|\mathbf{X}_{-i})]$ where $\psi(\bullet)$ is the operator taking the absolute local variation.

Therefore, μ^* is the best parallel to S_{T_i} as far as the operator ψ can be assimilated into the variance operator.

The Morris design focuses on the problem of sampling a number, r, of elementary effects from each distribution F_i (and hence from each G_i) in order to estimate the distribution's statistics. In the simplest form, since each elementary effect requires the evaluation of y twice, the total computational effort required for a random sample of r values from each F_i, is $n = 2rk$ runs, where k is the number of input factors. The *economy* of the design, defined by Morris as the number of elementary effects produced by the design divided by the number of experimental runs necessary to produce them, is then $rk/2rk$, i.e. $1/2$.

Morris suggests a more efficient design, with a larger value of the economy. Note that the larger the value of the economy for a particular design or method, the better it is in terms of providing information for sensitivity. The design proposed by Morris is based on the construction of a matrix, $\mathbf{B}^*$, of dimension k-by-$(k+1)$, whose rows represent input vectors $\mathbf{x}$s, for which the corresponding experiment provides k elementary effects, one for each input factor, from $(k+1)$ runs. The economy of the design is therefore increased to $k/(k+1)$.

A convenient choice for the parameters p and Δ of the design is p even and Δ equal to $p/[2(p-1)]$. This choice has the advantage that, although the design sampling strategy does not guarantee equal-probability sampling from each F_i, at least a certain symmetric treatment of inputs that may be desirable is ensured (for details see Morris (1991)).

The Morris designs starts by randomly selecting a 'base' value $\mathbf{x}^*$ for the vector $\mathbf{X}$. Each component x_i of $\mathbf{x}^*$ is sampled from the set $\{0, 1/(p-1), 2/(p-1), \ldots, 1\}$. Note that the vector $\mathbf{x}^*$ is used to generate the other sampling points but it is not one of them. The model is never evaluated at $\mathbf{x}^*$. The first sampling point, $\mathbf{x}^{(1)}$, is obtained by increasing one or more components of $\mathbf{x}^*$ by Δ. The choice of the components of $\mathbf{x}^*$ to be increased is conditioned by $\mathbf{x}^{(1)}$ still being in Ω. The second sampling point is generated from $\mathbf{x}^*$ with the property that it differs from $\mathbf{x}^{(1)}$ in its ith component that has been either increased or decreased by Δ. The index i is randomly selected in the set $\{1, 2, \ldots, k\}$. In mathematical notation $\mathbf{x}^{(2)} = (x_1^{(1)}, \ldots, x_{i-1}^{(1)}, x_i^{(1)} \pm \Delta, x_{i+1}^{(1)}, \ldots,$ $x_k^{(1)}) = (\mathbf{x}^{(1)} \pm \mathbf{e}_i \Delta)$. The third sampling point, $\mathbf{x}^{(3)}$, is again generated from the 'base' value $\mathbf{x}^*$. One or more of the k components of $\mathbf{x}^*$ are increased by Δ, with the property that $\mathbf{x}^{(3)}$ differs from $\mathbf{x}^{(2)}$ for only one component j, for any $j \neq i$. It can be either $x_j^{(3)} = x_j^{(2)} + \Delta$ or $x_j^{(3)} = x_j^{(2)} - \Delta$. The design proceeds producing a succession of $(k+1)$ sampling points $\mathbf{x}^{(1)}, \mathbf{x}^{(2)}, \ldots, \mathbf{x}^{(k+1)}$, with the key property that two consecutive points differ in only one component. Furthermore any component i of the 'base vector' $\mathbf{x}^*$ has been selected at least once to be increased by Δ in order to calculate one elementary effect for each factor.

Note that while each component of the 'base' vector $\mathbf{x}^*$ can only be increased (and not decreased) by Δ, a sampling point $\mathbf{x}^{(l+1)}$, with l in $\{1, \ldots, k\}$, may be different from $\mathbf{x}^{(l)}$ also because one of its components has been decreased (see example below).

The succession of sampling points $\mathbf{x}^{(1)}, \mathbf{x}^{(2)}, \ldots, \mathbf{x}^{(k+1)}$ defines what is called a *trajectory* in the input space. It also defines a matrix $\mathbf{B}^*$, with dimension $(k+1) \times k$, whose rows are the vectors $\mathbf{x}^{(1)}, \mathbf{x}^{(2)}, \ldots, \mathbf{x}^{(k+1)}$. $\mathbf{B}^*$ represents the design matrix and is called the Orientation matrix. An example of a trajectory is given in Figure 4.2 for $k = 3$.

Once a trajectory has been constructed and the model evaluated at its points, an elementary effect for each factor i, $i = 1, \ldots, k$, can be computed. If $\mathbf{x}^{(l)}$ and $\mathbf{x}^{(l+1)}$, with l in the set $\{1, \ldots, k\}$, are two

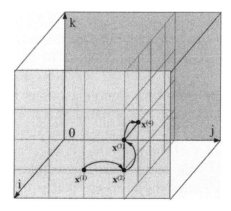

Figure 4.2   An example of trajectory in the input factor space when $k = 3$

sampling points differing in their ith component, the elementary effect associated with the factor i is either

$$d_i(\mathbf{x}^{(l)}) = \frac{[y(\mathbf{x}^{(l+1)}) - y(\mathbf{x}^{(l)})]}{\Delta}, \tag{4.2}$$

if the ith component of $\mathbf{x}^{(l)}$ has been increased by Δ or

$$d_i(\mathbf{x}^{(l)}) = \frac{[y(\mathbf{x}^{(l)}) - y(\mathbf{x}^{(l+1)})]}{\Delta}, \tag{4.3}$$

if the ith component of $\mathbf{x}^{(l)}$ has been decreased by Δ.

In other words, the orientation matrix $\mathbf{B}^*$ provides a single elementary effect per input factor and corresponds to a trajectory of k steps, in the input space, with starting point $\mathbf{x}^{(1)}$. Technicalities on how to build a design orientation matrix are given in Section 4.3.

The goal of the experiment is to estimate the mean and the variance of the distributions F_i and $G_i, i = 1, \ldots, k$. To this end a random sample of r elements from each F_i has to be selected, thus automatically providing a corresponding sample of r elements belonging to G_i. The extraction of such a sample requires the construction of r orientation matrices, independently generated, corresponding to r different trajectories in the input space. Each trajectory has a different starting point that is randomly generated. Since each orientation matrix provides an elementary effect per factor, the r matrices all together provide k r-dimensional samples, one for each F_i.

Although a characteristic of this sampling method is that points belonging to the same trajectory are not independent, the r points sampled from each F_i belong to different trajectories and are therefore independent. The same obviously applies to G_i. Therefore, the mean and standard deviation of each distribution F_i and G_i can be estimated by using the same estimators that would be used with independent random samples, i.e. as

$$\mu = \sum_{i=1}^{r} d_i/r \tag{4.4}$$

$$\sigma = \sqrt{\sum_{i=1}^{r}(d_i - \mu)^2/r} \tag{4.5}$$

where d_i, $i = 1, \ldots, r$, are the r elementary effects (or their absolute values) sampled from F_i (or from G_i).

Results of the Morris experiment can be easily interpreted. A large (absolute) measure of central tendency for G_i, i.e. a value of mean that is substantially different from zero, indicates an input with an important 'overall' influence on the output.

A large measure of spread, i.e. a high value of the standard deviation of F_i, indicates an input with a non-linear effect on the output, or an input involved in interaction with other factors. To rank factors in order of importance it is advisable to use μ^*, as this measure provides an estimate of the overall factor importance.

4.3   Implementing the method

To implement the Morris design, a number, r, of orientation matrices $\mathbf{B}^*$ have to be constructed. To build a matrix $\mathbf{B}^*$, the first step is the selection of a matrix $\mathbf{B}$, whose dimensions are $(k + 1) \times k$, with elements that are 0s and 1s and the key property that for every column index j, $j = 1, \ldots, k$, there are two rows of $\mathbf{B}$ that differ only in the jth entry. A convenient choice for $\mathbf{B}$ is a strictly lower triangular matrix of 1s.

The matrix $\mathbf{B}'$, given by,

$$\mathbf{B}' = \mathbf{J}_{k+1,1}\mathbf{x}^* + \Delta\mathbf{B}, \tag{4.6}$$

where $\mathbf{J}_{k+1,k}$ is a $(k + 1) \times k$ matrix of 1s, and $\mathbf{x}^*$ is a randomly chosen 'base value' of $\mathbf{X}$, could be used as a design matrix, since the corresponding experiment would provide k elementary effects, one for each input factor, with a computational cost of $(k + 1)$ runs. However, the problem with $\mathbf{B}'$ is that the k elementary effects that it produces would not be randomly selected.

Assume that $\mathbf{D}^*$ is a k-dimensional diagonal matrix in which each element is either $+1$ or -1 with equal probability, and $\mathbf{P}^*$ is a k-by-k random permutation matrix in which each column contains one element equal to 1 and all others equal to 0 and no two columns have 1s in the same position.

A randomized version of the sampling matrix is given by

$$\mathbf{B}^* = (\mathbf{J}_{k+1,1}\,\mathbf{x}^* + (\Delta/2)[(2\mathbf{B} - \mathbf{J}_{k+1,k})\mathbf{D}^* + \mathbf{J}_{k+1,k}])\mathbf{P}^*. \quad (4.7)$$

$\mathbf{B}^*$ provides one elementary effect per input, which is randomly selected.

Example

Consider a model with two input factors taking values in the set $\{0, 1/3, 2/3, 1\}$. In this case $k = 2$, $p = 4$, and $\Delta = 2/3$.

The matrix $\mathbf{B}$ is given by

$$\mathbf{B} = \begin{bmatrix} 0 & 0 \\ 1 & 0 \\ 1 & 1 \end{bmatrix},$$

and the randomly generated $\mathbf{x}^*$, $\mathbf{D}^*$ and $\mathbf{P}^*$ are

$$\mathbf{x}^* = (0, 1/3); \ \mathbf{D}^* = \begin{bmatrix} 1 & 0 \\ 0 & -1 \end{bmatrix}; \ \mathbf{P}^* = \mathbf{I}.$$

For these values, then

$$(\Delta/2)[(2\mathbf{B} - \mathbf{J}_{k+1,k})\mathbf{D}^* + \mathbf{J}_{k+1,k}] = \begin{bmatrix} 0 & \Delta \\ \Delta & \Delta \\ \Delta & 0 \end{bmatrix} = \begin{bmatrix} 0 & 2/3 \\ 2/3 & 2/3 \\ 2/3 & 0 \end{bmatrix}$$

and

$$\mathbf{B}^* = \begin{bmatrix} 0 & 1 \\ 2/3 & 1 \\ 2/3 & 1/3 \end{bmatrix},$$

so that

$$\mathbf{x}^{(1)} = (0, 1); \ \mathbf{x}^{(2)} = (2/3, 1); \ \mathbf{x}^{(3)} = (2/3, 1/3).$$

Figure 4.3 shows the resulting trajectory in the input space.

When implementing the Morris exercise on a model, the first problem to be addressed concerns the choice of the p levels among which each input factor is varied. For a factor following a uniform distribution, the levels are simply obtained by dividing the interval in which each factor varies into equal parts. For a factor following

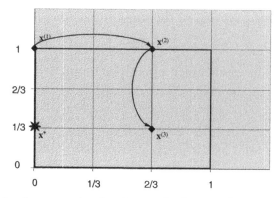

Figure 4.3   An example of trajectory in the two-dimensional space.

distributions other than uniform, it is opportune to select the levels in the space of the quantiles of the distribution.

Input values are then not sampled directly. Instead, the sampling is carried out in the space of the quantiles of the distributions, which is a k-dimensional hyper-cube (each quantile varies in $[0, 1]$). Then, given a quantile value for a given input factor, the actual value taken by the factor is derived from its known statistical distribution (Campolongo *et al.*, 1999).

The choice of the number of levels, p, or, in other words, the choice of the sampling step Δ, which is linked to p by the relation $\Delta = p/2(p-1)$, is an open problem. The choice of p is strictly linked to the choice of r. When the sampling size r is small, it is likely that not all the possible factor levels are explored within the experiment. For instance, in the above example, if $r = 1$, factor 1 never gets the values $1/3$ and 1 while factor 2 never gets the values 0 and $2/3$. Increasing the sampling size, thus reproducing the matrix $\mathbf{B}^*$ r times, would increase the probability that all the levels are explored at least once. Considering a high value of p, thus producing a high number of possible levels to be explored, only appears to augment the accuracy of the sampling. If this is not coupled with the choice of a high value of r, the effort will be wasted as many possible levels will remain unexplored. Previous experiments (Campolongo and Saltelli, 1997; Campolongo *et al.*, 1999; Saltelli *et al.*, 2000b, p. 367) have demonstrated that the choice of $p = 4$ and $r = 10$ has produced valuable results. Morris (1991) used a sample size of $r = 4$, this is probably the minimum value to place confidence in the experiment results.

The Morris method is implemented in SIMLAB.

4.4 Putting the method to work: an analytical example

The twenty factor analytical example used to test the performance of the Morris method, which is taken from Morris (1991), was described in Section 3.7. Parameters of the Morris experiment were set respectively to $p = 4$, $\Delta = 2/3$ and $r = 4$. Using the same representation as in Morris (1991), the values obtained for the sensitivity measures μ and σ are displayed in Figure 4.4.

The pattern described in Figure 4.4 almost reproduces the one shown in Figure 1 of Morris (1991). Input variables 1–10, which are supposed to have a significant effect on the output, are well separated from the others. In particular, as shown in Morris (1991), variables 8, 9 and 10 are separated from the others because of their high mean (abscissa) values. Hence, considering both means and standard deviations together, one can conclude that the first ten factors are important; of these, the first seven have significant effects that involve either interactions or curvatures; the other three are important mainly because of their first-order effect.

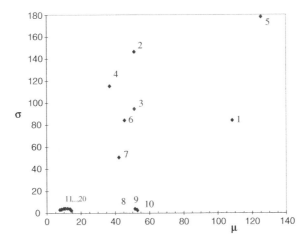

Figure 4.4 Results of the Morris experiment on the analytical model described in Chapter 3.

4.5   Putting the method to work: sensitivity analysis of a fish population model

The method of Morris in its extended version is applied to the model of fish population dynamics described in Chapter 3. The goal of the experiment is twofold: (i) to establish the relative importance of the various physical and ecological process involved in the dynamics of the fish population, so improving our understanding of the system; and/or (ii) to eliminate those factors or group of factors that seem to be irrelevant in order to reduce the complexity of the model and increase its efficiency.

To start the analysis, the quantities of interest have to be determined. First, it is essential to specify which is the model response (or model output) regarded as the most informative for the goal of the analysis. In this study, we focus on λ_{max}, which is the dominant eigenvalue of the population matrix. The eigenvalue λ_{max} represents the population growth rate. If $\lambda_{max} = 1$ the population is stationary. In particular, the quantity of interest is λ^{365}, which is λ_{max} after one year simulation time, representing the annual population growth. The total number of model input factors is 103. Of these, 72 are factors that represent the daily natural mortality (Z), duration (D), and daily fecundity (F) of each of the three species under study (sardines (I), anchovies (J) and mackerel (K)). Their best values and ranges of variation are specified in the Tables 3.2–3.4 in Chapter 3. Best values are chosen to produce a dominant eigenvalue in the population matrix equal to 1. Daily fecundity factors of early development stages with min = max = 0 are not considered.

To simplify the notation, these 72 factors are denoted by two capital letters, the first indicating the type of factor (Z, D or F), and the second indicating the species to which it is referring (I, J or K). The numbers between brackets denote the life-stage: $i = 1, 13$ for sardines, i.e. from egg to late adult following the life stages given in Table 3.2, and $i = 1, 9$ for anchovies and mackerel, from egg to late adult following the life stages given respectively in Tables 3.3 and 3.4. For example, ZJ(3) denotes the mortality (Z) of anchovies (J) in the early larvae stage. The remaining 31 inputs are factors

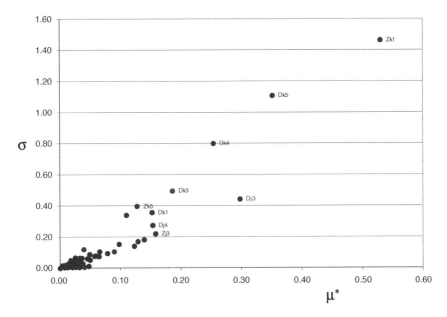

Figure 4.5 Graph displaying the Morris sensitivity measures μ^* and σ for the 103 model input factors. Only the most important factors are labelled.

involved in the migration and interspecies competition between larvae and juveniles. Lower case letters denote these factors.

The extended Morris method is applied to the fish population model with a sample size $r = 10$. Each of the 103 input factors is assumed to follow a uniform distribution between its extreme values, reported in Tables 3.2–3.4. In the design, each factor is varied across four levels ($p = 4$). A total number $N = 1040$ of model evaluations is performed ($N = r \times (k + 1)$, where k is the number of input factors).

The results of the experiment exercise are shown in Figure 4.5, where the sensitivity measures μ^* and σ are plotted for the 103 input factors. Labels indicating the names of the factors are given only for the ten most important factors.

Factors can also be ranked in (decreasing) order of importance according to μ^*, which is a measure of the overall factor importance. However, it is worth noting that in this case the order of importance that would have been obtained by using σ instead of

μ^* is very similar. Results of the experiment show that each input factor with a high value for the estimated mean, μ^*, also has a high value for the estimated standard deviation, σ, or in other words, none of the factors has a purely linear effect. This is also evident from Figure 4.5, where all the points lie around the diagonal.

A number of conclusions can be drawn by examining Figure 4.5. The first group of factors is clearly separated from the others, indicating a large influence on the population growth rate. These are (in decreasing order of importance): ZK(1), the daily natural mortality for mackerel at the egg stage; DK(5) and DK(4), the duration for mackerel at the juvenile and early juvenile stage respectively; DJ(3), the duration for anchovies at the early larvae stage, which is also the factor that is less involved in the interaction and/or curvature effects, as it does not lie exactly on the diagonal of the (μ^*, σ) plane but in the $\mu^* > \sigma$ zone. A second group of factors that are quite influential on the output include (not in order of importance): DJ(4), DJ(6) and DJ(7), i.e. the duration for anchovies at the late larvae, late juvenile and prerecruit stages; DK(1), DK(2) and DK(3), i.e. the duration for mackerel at the egg and early and late larvae stages; ZJ(1), ZJ(3), ZJ(4), ZJ(5) and ZJ(6), i.e. the mortality for anchovies at the egg, early and late larvae, and early and late juvenile stages; and ZK(5), i.e. the mortality for mackerel at the juvenile stage. Then other factors follow immediately: DJ(2), i.e. the duration for anchovies at the yolk-sac larvae stage, ZJ(7), i.e. the mortality for anchovies at the prerecruit stage, and so on. Values of μ^* for these remaining factors decrease smoothly, without any discontinuity, indicating that it is very difficult to distinguish a group of important factors from a group of non-important ones.

Input factors involved in interspecies competition between larvae and juveniles are not very important with respect to the others. None of them appear among the first thirty that were identified by Morris, even if they were responsible for the fluctuations observed in the simulations. This means they do not influence the magnitude of the fluctuations. The conclusion is that, if our modelling approach is correct, it would be difficult to identify interspecies competition in real time series data, as it will be masked by environmental fluctuations.

Other relevant conclusions include:

1. the daily fecundity (F) factors are not very significant for any of the three species at any life stage;

2. none of the most important twenty factors are related to adult life stages;

3. the daily natural mortality and duration factors play a more substantial role on the dynamics of the three populations;

4. the sardine population appears to be less influential on the overall population dynamics than do other populations.

The results of the screening experiment have, on the one hand, contributed to improving our understanding of the fish population dynamics; on the other hand, they may be used to update the model in order to make it more efficient and/or more consistent with observed data. For instance, point (4) above may call for a revision of the role of sardines in the model or one may think of focusing on a simplified version of the model obtained by eliminating the parameters relative to interspecies competition between larvae and juveniles that do not seem to play a substantial role. Furthermore, results can be used to prioritise further research and experiments by addressing the estimates of those parameters that have the greatest effect on the output of interest.

4.6 Conclusions

We suggest that one uses the method discussed in this chapter when the problem is that of screening a few important input factors among a large number contained in the model, or in other words for setting Factor's Fixing. Operatively, when factors with small values of μ^* are identified, these can be fixed at any value in their uncertainty distribution without any significant loss of information. At times, this may lead to segment of the model being dropped.

The main advantage of the method is its low computational cost: the plan then requires a total number of runs that is a linear function of the number of examined factors.

The method has a number of advantages with respect to other screening methods that are widely accepted in the literature. For example, with respect to methods based on Fractional Factorials (Saltelli *et al.*, 2000b, p. 53), the Morris design is computationally more efficient. A fractional factorial design with high resolution (high resolution is desirable to avoid confounding effects of different orders) may be too expensive to perform.

A more economical design such as that proposed by Cotter (1979) relies on strict assumptions and fails when these assumptions are not fulfilled.

The Iterated Fractional Factorial Design proposed by Andres and Hajas (1993) is based on the idea of grouping factors and performs appropriately when the number of factors that is important is restricted. The sequential bifurcation proposed by Bettonvil (1990) is applicable only when factor effects have *known signs*, which means the analyst knows whether a specific individual factor has a positive or negative effect on the simulation response, an assumption that is rarely fulfilled.

The Morris method, as all screening methods, provide sensitivity measures that tend to be qualitative, i.e. capable of ranking the input factors in order of importance, but do not attempt to quantify by how much one given factor is more important than another. A quantitative method would provide an estimate, for example, of the exact percentage of total output variance that each factor (or group of factors) accounts for. However, quantitative methods are more computational expensive (see Chapter 5) and not affordable when a large number of input factors are involved in the analysis or the model is time consuming.

5 METHODS BASED ON DECOMPOSING THE VARIANCE OF THE OUTPUT

The purpose of this chapter is to describe in some detail the variance based methods that were succinctly introduced in Chapter 1. We start by showing how they can tackle some of the settings discussed in Chapter 2, and how their properties compare with respect to what one might desire, from model independence to the capacity to assess the importance of groups of uncertain factors (Chapter 2, Table 2.1). At the end of this chapter, we suggest strategies for the estimation of the sensitivity measures for the two radically different cases of orthogonal and non-orthogonal input factors.

5.1 The settings

We recall briefly the settings for sensitivity analysis defined in Chapter 2.

1. In *Factors Prioritisation (FP) Setting* we want to make a rational bet on what is the factor that one should fix to achieve the greatest reduction in the uncertainty of the output.

2. In *Factors Fixing (FF) Setting* we try to screen the input factors by identifying factors or sets of factors that are non-influential.

3. In *Variance Cutting (VC) Setting* we would like to make a rational bet on what is the minimal subset of factors that one should fix to achieve a prescribed reduction in the uncertainty of the output.

4. In *Factors Mapping (FM) Setting* we look for factors mostly responsible for producing realisations of Y in a given region.

Sensitivity Analysis in Practice: A Guide to Assessing Scientific Models A. Saltelli, S. Tarantola, F. Campolongo and M. Ratto © 2004 John Wiley & Sons, Ltd. ISBN 0-470-87093-1

As we shall see below, the first three settings are particularly apt to structure our analysis of model sensitivity in terms of variance. In turn, they also offer a quite intuitive way to interpret the sensitivity measures that are 'variance based'. The fourth setting is mostly related to Monte Carlo filtering techniques and will be analysed in detail in Chapter 6.

5.2   Factors Prioritisation Setting

We shall discuss here Setting FP in some detail, with examples, to offer the reader an overview of what it implies and how it can be tackled. Some of the key concepts have already been touched upon in Chapters 1 and 2, but they are repeated here in full.

Let us assume that all the factors X are left free to vary over their entire range of uncertainty. The corresponding uncertainty of the model output $y = f(X)$ is quantified by its unconditional variance $V(Y)$.

Our objective in this setting is to rank the factors according to the amount of output variance that is removed when we learn the true value of a given input factor X_i.

The factors could then be ranked according to $V(Y|X_i = x_i^*)$, the variance obtained by fixing X_i to its true value x_i^*. This variance is taken over all factors but X_i. We could normalise it by the output (unconditional) variance, to obtain $V(Y|X_i = x_i^*)/V(Y)$. Note that $V(Y|X_i = x_i^*)$ could even be larger than $V(Y)$ for particular values of x_i^* (see Box 2.4 Conditional and unconditional variances, Chapter 2). The problem is that we do not know what x_i^* is for each X_i. It hence sounds sensible to look at the average of the above measure over all possible values x_i^* of X_i, i.e. $E(V(Y|X_i))$, and take the factor with the smallest $E(V(Y|X_i))$. We have dropped the dependence from x_i^* in the inner variance as this is eliminated by the outer mean. In a richer notation, to indicate the conditioning argument, we could write it as $E_{X_i}(V_{X_{-i}}(Y|X_i))$ where X_{-i} denotes the vector of all the input factors but factor X_i.

Given that $V(Y)$ is a constant and $V(Y) = V(E(Y|X_i)) + E(V(Y|X_i))$, betting on the lowest $E(V(Y|X_i))$ is equivalent to betting on the highest $V(E(Y|X_i))$.

Several practitioners of sensitivity analysis have come up with different estimates of $V_i = V(E(Y|X_i))$ as a measure of sensitivity. Investigators have given the V_i, or the ratios $S_i = V_i/V(Y)$, the names 'importance measure', 'correlation ratio' or 'sensitivity index' (see Saltelli *et al.*, 2000a, p. 167 and Saltelli *et al.*, 1999, 2000b for reviews). Statisticians and practitioners of experimental design call V_i (or S_i) the 'first-order effect' of X_i on Y (see, for example, Box *et al.*, 1978).

Based on this discussion, we have already claimed in Chapter 1 that the measure S_i is the proper instrument to use for Setting FP. In the next sections we illustrate this with simple examples, while recalling some concepts associated with the variance based methods.

5.3   First-order effects and interactions

We need to have a few words here about the concept of 'interaction', repeating some of the concepts introduced in Chapter 1. Two factors are said to interact when their effect on Y cannot be expressed as a sum of their single effects on Y. This definition will become more precise in the following. Interactions may imply, for instance, that extreme values of the output Y are uniquely associated with particular combinations of model inputs, in a way that is not described by the first-order effects S_i, just mentioned. Interactions represent important features of models, e.g. when models are employed in risk analysis. Interactions are also more difficult to detect than first-order effects. For example, by using regression analysis tools it is fairly easy to estimate first-order indices, but not interactions (remember the relationship $S_{x_i} = \beta_{x_i}^2$ discussed in Chapter 1 for linear models and orthogonal inputs, where β_{x_i} is the standardised regression coefficient for factor X_i). Interactions are described well in terms of variance. For example, the effect of the interaction between two orthogonal factors X_i and X_r on the output Y can be defined in terms of conditional variances as:

$$V_{ir} = V(E(Y|X_i, X_r)) - V(E(Y|X_i)) - V(E(Y|X_r)). \qquad (5.1)$$

In this equation, $V(E(Y|X_i, X_r))$ measures the joint effect of the pair (X_i, X_r) on Y. In $V(E(Y|X_i, X_r))$ the inner average is evaluated over the space of all factors but X_i, X_r and the outer variance over all possible values of X_i, X_r.

The term V_{ir} is the joint effect of X_i and X_r minus the first-order effects for the same factors. This is known as a second-order, or two-way, effect (Box *et al.* 1978). Analogous formulae can be written for higher orders as we shall see in a moment.

When the input factors are orthogonal, the conditional variances such as $V(E(Y|X_i))$ and $V(E(Y|X_i, X_r))$ can be seen in the context of the general variance decomposition scheme proposed by Sobol' (1990), whereby the total output variance $V(Y)$ for a model with k input factors can be decomposed as:

$$V(Y) = \sum_i V_i + \sum_i \sum_{j>i} V_{ij} + \ldots + V_{12\ldots k} \qquad (5.2)$$

where

$$V_i = V(E(Y|X_i)) \qquad (5.3)$$

$$V_{ij} = V(E(Y|X_i, X_j)) - V_i - V_j \qquad (5.4)$$

$$V_{ijm} = V(E(Y|X_i, X_j, X_m)) - V_{ij} - V_{im} - V_{jm} - V_i - V_j - V_m \quad (5.5)$$

and so on. A model without interactions, i.e. a model for which only the terms in (5.3) are different from 0, is said to be additive in its factors.

5.4 Application of S_i to Setting 'Factors Prioritisation'

Imagine that the set of the input factors $(X_1, X_2, \ldots, X_k)$ is orthogonal. If the model is additive, i.e. if the model does not include interactions between the input factors, then the first-order conditional variances $V_i = V(E(Y|X_i))$ are indeed all that we need to know in order to decompose the model's variance. In fact, for additive models,

$$\sum_i V_i = V(Y) \qquad (5.6)$$

or, equivalently,

$$\sum_i S_i = 1. \qquad (5.7)$$

Let us now assume that the influence of factors X_i and X_r on the output Y, which is measured by V_i and V_r, is smaller than the influence of the factor X_j, which is quantified by V_j. Imagine also that the output Y is influenced by an interaction (or synergy) between X_i and X_r. Are we now sure that X_j is more important than X_i and X_r under Setting FP? Or should the answer depend upon the interaction between X_i and X_r on Y? It should not, as no assumption of additivity was made in the discussion of Section 5.2 above. To confirm this let us look at a didactical example: a simple function $Y = m(X_1, X_2)$ where one of the two S_is is zero and a large interaction component is present. This exercise will also help us to get accustomed to the role of interactions in models.

A Legendre polynomial of order d is discussed in McKay (1996):

$$Y = L_d(X) = \tfrac{1}{2^d} \sum_{m=0}^{d/2} (-1)^m \binom{d}{m} \binom{2d - 2m}{d} X^{d-2m} \quad (5.8)$$

where $L_d(X)$ is a function of two orthogonal input factors, X (an independent variable) and d (the order of the polynomial). X is a continuous uniformly distributed factor in $[-1, 1]$, and d is a discrete uniformly distributed factor in $d \in [1, 5]$. $L_d(X)$ is shown in Figure 5.1.

Consider the decomposition (5.2) for the model $L_d(X)$: $1 = S_X + S_d + S_{Xd}$. Only two of these terms are non-zero: $S_X = 0.2$ and $S_{Xd} = 0.8$, while $S_d = 0$. The equality $S_d = 0$ can easily be explained (Figure 5.1). Fixing d at any given value does not help reduce the uncertainty in $L_d(X)$. This is because the mean of $L_d(X)$ over X does not depend on d in $d \in [1, 5]$. On the other hand, the best gain that can be expected in terms of reduction of output variance is 20%, that obtained on average by learning X. We recall that $S_X = \{V(Y) - E[V(Y|X)]\}/V(Y) = 0.2$. If setting FP is phrased in terms of a bet on which factor will reduce the output variance the most, then this case tells us that the best choice is to bet on factor X. For orthogonal factors, the proper measure of sensitivity to use in order to rank the input factors in order of importance according to setting FP is V_i, or equivalently $S_i = V_i / V(Y)$, whether or not the factors interact.

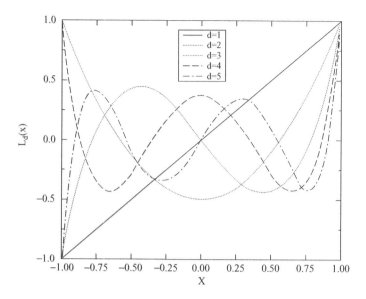

Figure 5.1   Legendre polynomials of order d.

What happens when the input factors are not independent of each other?[1] We can anticipate here that the same conclusion applies, i.e. the first-order indices S_i remain the measure to use for setting FP. Because interactions (a property of the model) and non-orthogonality (a property of the factors) can often interplay, it is instructive to consider some simple examples to see these effects at work.

Consider a linear and additive model (*Model 1*):

$$y = \sum_i X_i \qquad (5.9)$$

and *Model 2*, a non-linear, non-additive model:

$$y = \sum_i X_i + 5 X_1 X_3 \qquad (5.10)$$

$$\mathbf{X} \equiv (X_1, X_2, X_3). \qquad (5.11)$$

---

[1] The most intuitive type of dependency among input factors is given by correlation. However, dependency is a more general concept than correlation, i.e. independency means orthogonality and also implies that correlation is null, while the converse is not true, i.e. null correlation does not necessarily imply orthogonality (see e.g. Figure 6.6 and the comments to it). The equivalence between null correlation and independency holds only for multivariate normal distributions.

To analyse non-orthogonality for both models we shall test three alternative correlation structures: $\mathbf{X} \sim N(\mathbf{0}, \mathbf{I})$, $\mathbf{X} \sim N(\mathbf{0}, \mathbf{C_1})$ and $\mathbf{X} \sim N(\mathbf{0}, \mathbf{C_2})$, where $\mathbf{I}$ is the identity matrix and:

$$\mathbf{C_1} = \begin{vmatrix} 1 & 0.7 & 0.1 \\ & 1 & 0.7 \\ & & 1 \end{vmatrix} \tag{5.12}$$

and

$$\mathbf{C_2} = \begin{vmatrix} 1 & 0.7 & 0. \\ & 1 & 0. \\ & & 1 \end{vmatrix}. \tag{5.13}$$

Model 1, no correlation
The factors are uncorrelated and follow a standard normal distribution (i.e. they are orthogonal). As already discussed, the first-order terms are sufficient for assessing the importance of orthogonal factors. All the V_i are equal to 1 and the total variance V is 3 (all the results are given in Table 5.1). Each of the three factors has first-order importance V_i/V equal to $1/3$, and in Setting FP; no factor can be identified as most important.

Table 5.1 V_i for Models 1 and 2 and different correlation structures.

	X_i	V_i	
		Model 1 (additive)	Model 2 (non additive)
	1	1	1
Correlation matrix $\mathbf{I}$	2	1	1
	3	1	1
Total variance		$V = 3$	$V = 28$
	1	3.2	4
Correlation matrix $\mathbf{C_1}$	2	5.7	18
	3	3.2	4
Total variance		$V = 6$	$V = 32$
	1	2.9	2.9
Correlation matrix $\mathbf{C_2}$	2	2.9	2.9
	3	1.0	1.0
Total variance		$V = 4.4$	$V = 29.3$

Model 1, Correlation C_1
The pairs X_1, X_2, and X_2, X_3 are correlated and X_3 is weakly correlated to X_1. Now, the total variance V increases to 6, V_2 raises to 5.7, both V_1 and V_3 increase to 3.2. This is because the correlations in which X_2 is involved are stronger than those of X_1 and X_3. The symmetry in the model and in the correlation structures yields $V_1 = V_3$. The largest gain in variance reduction is obtained by fixing X_2. There are no reasons to question this statement, thus confirming that in Setting FP, also in the presence of correlations, the fractional variances V_i are the measure to choose to make an informed choice. What happens instead if the model is non-additive?

Model 2, no-correlation
The total variance is $V = 28$, and the fraction of output variance explained by the first-order terms is only $\sum_{i=1}^{3} V_i / V = 3/28$. The second-order term V_{13} accounts for the remaining $25/28$. Nothing new here; as for the case of the Legendre polynomial, we can base our analysis on the first-order terms V_i and, for the purpose of our bet (which factor would one fix to obtain the largest expected reduction in variance?), the three factors are to be considered equally important. There is only one case remaining, i.e. that of the simultaneous occurrence of correlation and interactions.

Model 2, correlation C_1
The total variance increases to $V = 32$, V_2 is equal to 18 and $V_1 = V_3 = 4$. The V_is identify X_2 as the most important factor. Note that:

1. Comparing the V_i for Models 1 (correlation C_1) and 2 (same correlation), we note that adding an interaction between X_1 and X_3 actually promotes X_2, i.e. increases its first-order effect. This promotion is indeed due to the correlation of X_2 with X_1 and X_3.

2. When $C \neq I$, it does not help to compare the summation of the first-order terms with the total variance. For Model 2 (correlation C_1), $\sum_i V_i = 26$, and $V = 32$, but we cannot conclude that the missing fraction of variance is that due to

interaction (Equation (5.2) does not hold here). The fraction due to interaction could be much higher.

Let us further consider *Models 1 and 2*, with the same correlation C_2, a case where only X_1 and X_2 are correlated.

The total variance for Model 1 becomes $V = 4.4$ and that for Model 2 becomes $V = 29.3$. We observe that:

1. Calculating the indices V_i, we obtain the same values for both Models 1 and 2, i.e. $V_1 = V_2 = 2.9$ and $V_3 = 1$. In other words, moving from Model 1 to Model 2 with the same correlation structure C_2, the first-order indices do not see the interaction between X_1 and X_3. This is because X_1 and X_2 are correlated but do not interact, while X_1 and X_3 interact but are not correlated.

2. Further, comparing Model 2, no-correlation, with the same model and correlation C_2, we see again that, for the same reasons as above, the effect of the interaction X_1, X_3 is not seen: $V_3 = 1$ for both cases.

But this is the same situation met in the Legendre polynomials' example. In fact it does not matter whether the factors interact or not, because under the Setting FP we indeed 'act' on one factor at a time, looking for the factor that, when fixed to its true – albeit unknown, value, guarantees, on average, the largest variance reduction of Y.

This measure is nothing other than V_i. With a given correlation structure $C \neq I$, the presence of interactions may ($C = C_1$) or may not ($C = C_2$) change the values of V_i and, hence, change the outcome of our informed choice. When it does not, this simply means that the factors involved do not gain or lose importance in this setting because of their interaction.

In other words, Setting 'Factors Prioritisation' assumes that factors are fixed singularly, as discussed in Section 5.2 above. This would normally prevent the detection of interactions. Yet in the presence of non-orthogonal input factors, fixing one factor also influences the distribution of the others. This may allow the influence of interactions to emerge, depending on the relative patterns of non-orthogonality (e.g. correlation) and interactions. In conclusion, in the FP Setting, the fractional variances V_i are the proper measure to use for making an informed choice, for any combination of interaction and non-orthogonality among factors.

When adopting the Setting FP, we accept the risk of remaining ignorant about an important feature of the model object of the sensitivity analysis: the presence of interactions in the model.

5.5   More on variance decompositions

Let us now return to our variance decomposition (5.2) (Sobol' 1990) which is valid for the orthogonal case, i.e.

$$V(Y) = \sum_i V_i + \sum_i \sum_{j>i} V_{ij} + \ldots + V_{12\ldots k},$$

where $V_i = V(E(Y|X_i))$, $V_{ij} = V(E(Y|X_i, X_j)) - V_i - V_j$ and so on. Sobol's decomposition is based on a decomposition of the function $Y = f(\mathbf{X})$ itself into terms of increasing dimensionality, i.e.,

$$f = f_0 + \sum_i f_i + \sum_i \sum_{j>i} f_{ij} + \ldots + f_{12\ldots k} \qquad (5.14)$$

where each term is a function only of the factors in its index, i.e. $f_i = f_i(X_i)$, $f_{ij} = f_{ij}(X_i, X_j)$ and so on. The decompositions in Equations (5.2) and (5.14) are unique, provided that the input factors are orthogonal and that the individual terms $f_{i_1 i_2 \ldots i_s}$ are square integrable over the domain of existence. Note that the first-order terms f_i in (5.14) are the same as those plotted in the Box 2.5 High dimensional model representation in Chapter 2.

As illustrated in Chapter 1, one important aspect of the Sobol' development is that similar decompositions can be written by taking the factors into subsets. Imagine that the factors have been partitioned into a trial set $\mathbf{u} = (X_{i_1}, X_{i_2}, \ldots, X_{i_m})$, and the remaining set $\mathbf{v} = (X_{l_1}, X_{l_2}, \ldots, X_{l_{k-m}})$. Then, according to Sobol', the total variance associated with $\mathbf{u}$ can be computed as

$$= V(Y) - V(E(Y|\mathbf{v})) \qquad (5.15)$$

In Equation (5.15) $V(E(Y|\mathbf{v}))$ is the first-order effect of the set $\mathbf{v}$, and $V(Y) - V(E(Y|\mathbf{u})) - V(E(Y|\mathbf{v}))$ is the interaction term between the sets $\mathbf{u}$ and $\mathbf{v}$. Equation (5.15) shows that the variance based sensitivity measure described in this chapter are indeed capable of treating factors in a group, as requested in Chapter 2.

We now introduce one last conditional variance (Homma and Saltelli, 1996), $V(E(Y|\mathbf{X}_{-j}))$. This is the closed contribution to the variance of Y due to non-X_j, i.e. to the $k-1$ remaining factors. This implies that, for orthogonal inputs, the difference $V(Y) - V(E(Y|\mathbf{X}_{-j}))$ is equal to the sum of all terms in the variance decomposition (Equation (5.2)) that include X_j. We illustrate this for the case $k = 3$:

$$S_{T1} = \frac{V(Y) - V(E(Y|\mathbf{X}_{-1}))}{V(Y)}$$

$$= \frac{E(V(Y|\mathbf{X}_{-1}))}{V(Y)} = S_1 + S_{12} + S_{13} + S_{123} \qquad (5.16)$$

and analogously:

$$\begin{aligned}
S_{T2} &= S_2 + S_{12} + S_{23} + S_{123} \\
S_{T3} &= S_3 + S_{13} + S_{23} + S_{123}
\end{aligned} \qquad (5.17)$$

where each sensitivity index is defined as $S_{i_1,\ldots,i_m} = V_{i_1,\ldots,i_m}/V(Y)$. The reader will remember the S_{Tj} as they have been introduced in Chapter 1. We have called the S_{Tj}s *total effect* terms. The total effects are useful for the purpose of SA, as discussed in Saltelli *et al.* (1999), as they give information on the non-additive part of the model. As mentioned, for a purely additive model and orthogonal inputs, $\sum_{i=1}^{k} S_i = 1$, while for a given factor X_j a significant difference between S_{Tj} and S_j flags an important role of interactions for that factor in Y.[2] Clearly the same information could be obtained by computing all terms in Equation (5.2), but these are as many as $2^k - 1$. This problem has been referred to as 'the curse of dimensionality'. For this reason we customarily tend to compute the set of all S_i plus the set of all S_{Ti}, which gives a fairly good description of the model sensitivities at a more reasonable cost.

---

[2] See the comments on Table 1.10 for the implications on main and total effects coming from non-orthogonal inputs.

Moreover, while the $S_{Tj} = [V(Y) - V(E(Y|\mathbf{X}_{-j}))]/V(Y)$ can always be computed, whatever the dependency structure among input factors, the decomposition (5.2) and the last right-hand side terms in Equations (5.16)–(5.17) are valid only for orthogonal inputs. Many applications of this strategy to different models can be found in various chapters of Saltelli *et al.* (2000a).

5.6 Factors Fixing (FF) Setting

One interesting remark about the total effect indices, S_{Tj}s, and the associated variances $E(V(Y|\mathbf{X}_{-j}))$ is that these give us the fraction of the total variance that would be left on average if all factors but X_j were fixed. In other words, $E(V(Y|\mathbf{X}_{-j}))$ represents the average output variance that would remain as long as X_j remains unknown. For these reason Jansen *et al.* (1994) call the $E(V(Y|\mathbf{X}_{-j}))$ 'bottom marginal variances'. They analogously call the $V(E(Y|X_j))$ 'top marginal variances'. The concept of bottom marginal could be turned into a setting: 'Which is the factor that, being left undetermined while all others are fixed, would leave the largest variance in the output?' The educated guess (because even here we do not know where the other factors are fixed) is: the factor with the highest $E(V(Y|\mathbf{X}_{-j}))$ or S_{Tj}. It is straightforward to understand that the S_{Tj}s are also the measure of choice to tackle the setting FF. If X_j has no influence at all, then fixing $\mathbf{X}_{-j}$ also fixes Y, and the measure $V(Y|\mathbf{X}_{-j})$ will equal zero. A fortiori, the average of $V(Y|\mathbf{X}_{-j})$ over X_j, i.e. $E(V(Y|\mathbf{X}_{-j}))$, will likewise be zero. On the other hand, if $E(V(Y|\mathbf{X}_{-j}))$ is zero, $V(Y|\mathbf{X}_{-j})$ must also be identically zero over the X_j axis, given that it cannot be negative. If this happens, then X_j is totally non-influential on Y. In conclusion, $S_{Tj} = 0$ is condition necessary and sufficient for X_j to be non-influential. Therefore, X_j can be fixed at any value within its range of uncertainty without affecting the value of the output unconditional variance $V(Y)$. This is an advantage as we can reduce the dimensionality of the input space.

In summary, the output variance $V(Y)$ can always be decomposed by conditioning with respect to both X_j and $\mathbf{X}_{-j}$, no matter whether the input is orthogonal or not:

$$V(Y) = V(E(Y|X_j)) + E(V(Y|X_j))$$
$$V(Y) = V(E(Y|\mathbf{X}_{-j})) + E(V(Y|\mathbf{X}_{-j})).$$

Let us normalise the two decompositions as:

$$\boxed{S_j} \leftarrow 1 = \boxed{\frac{V\!\left(E\!\left(Y|X_j\right)\right)}{V(Y)}} + \frac{E\!\left(V\!\left(Y|X_j\right)\right)}{V(Y)}$$

$$1 = \frac{V\!\left(E\!\left(Y|\mathbf{X}_{-j}\right)\right)}{V(Y)} + \boxed{\frac{E\!\left(V\!\left(Y|\mathbf{X}_{-j}\right)\right)}{V(Y)}} \rightarrow \boxed{S_{Tj}}.$$

From the first term in the first decomposition we can obtain S_j, whilst from the second term in the second decomposition we obtain S_{Tj}.

5.7 Variance Cutting (VC) Setting

As has been done already in Chapter 1, we would like to introduce a more compact notation for the higher-order conditional variances (e.g. Equations (5.3)–(5.5)). Let us use $V^c_{i_1 i_2 \ldots i_s}$, where the superscript c stands for 'closed', to indicate the sum of all $V_{i_1 i_2 \ldots i_s}$ terms in Equation (5.2) that is closed in the indices $i_1, i_2, \ldots i_s$. This is expressed as $V^c_{i_1 i_2 \ldots i_s} = V(E(Y|X_{i_1}, X_{i_2}, \ldots, X_{i_s}))$ giving, for example,

$$V^c_i = V_i = V(E(Y|X_i)) \tag{5.18}$$

$$V^c_{ij} = [V_i + V_j + V_{ij}] = V(E(Y|X_i, X_j)) \tag{5.19}$$

$$V^c_{ijm} - [V_i + V_j + V_m + V_{ij} + V_{jm} + V_{im} + V_{ijm}]$$
$$= V(E(Y|X_i, X_j, X_m)) \tag{5.20}$$

and so on, where the expressions in square brackets hold only for orthogonal inputs. Likewise $V^c_{-i_1 i_2 \ldots i_s}$ will indicate the sum of all $V_{l_1 l_2 \ldots l_{k-s}}$ that are closed within the complementary set of $i_1, i_2, \ldots i_s$.

We are now ready to suggest a strategy to tackle the problem Setting VC. This is particularly complex, especially for the general case where the input factors are non-orthogonal. The problem with non-orthogonal input, in brief, is that the reduction in variance that can be achieved fixing one factor depends on whether or not other factors have been fixed, and the incremental reduction in variance for each factor depends on the order in which the factors are fixed. Equation (5.2) loses its uniqueness in this case. One can still compute closed variances, such as $V_{ij}^c = V(E(Y|X_i X_j))$, but this can no longer be decomposed as first order and interaction effects in a unique way. Imagine that the Setting 'Variance Cutting' as been formulated, again in terms of a bet, so that in order to win one has to identify the smallest subset $\mathbf{u} = (X_{i_1}, X_{i_2}, \ldots X_{i_m})$ of $\mathbf{X}$ so that fixing $\mathbf{u}$, the variance of Y will be reduced by V_r, with $V_r < V$.

In Saltelli and Tarantola (2002), we have suggested the following empirical procedures for this Setting when the input factors are orthogonal. We compute the full set of V_js and V_{Tj}s and use the latter to rank the factors. A sequence $V_{TR_1}, V_{TR_2}, \ldots, V_{TR_k}$ is thus generated where $V_{TR_1} > V_{TR_2} > \ldots > V_{TR_k}$. If V_{R_1} is larger than V_r, then X_{R_1} is the factor on which we can bet, as it is more likely that, fixing X_{R_1} at some value within its range of uncertainty, we actually obtain $V_r < V$. If V_{R_1} is smaller than V_r, we have to consider the factor with the second highest total index, i.e. X_{R_2}, and we check whether $V_{R_1 R_2}^c > V_r$. If this happens, we will bet on the pair X_{R_1}, X_{R_2}. It this does not happen, we have to add X_{R_3} and so on. This procedure should work for additive as well as for non-additive models, and its empirical justification is that one seeks to fix the factors with the highest overall 'total effect' in the hope that the right combination of factors will yield the desired reduction. Clearly this procedure has an alternative in a brute force search of all combinations of factors in Equation (5.2). This search will certainly be more informative but it would again confront us with the curse of dimensionality. The procedure for the non-orthogonal case (additive or non-additive model) is more complex and the reader is referred to Saltelli and Tarantola (2002). As a trace, we can say that in this case an empirical approach is also taken, favouring the selection of factors with high overall interactions (as before), and trying not to select for the trial those factors with high average correlation with the factors already fixed.

In conclusion, we see that for the orthogonal case, a rational selection strategy for the subset of interest is based on the computation of the full sets of S_{T_j} (but when doing this all the S_j are computed as well, see the following sections), plus a variable number of higher-order conditional variances. This strategy is meant to fight the curse of dimensionality, as attempting all combination of factors in a brute-force search for the smallest subset of $\mathbf{X}$ that gives the desired reduction in $V(Y)$ would be computationally prohibitive; one would have to compute all $2^k - 1$ terms in Equation (5.2).

Note that, for the non-orthogonal case, one might still engage in a brute force search computing all possible closed terms $V^c_{i_1 i_2 \ldots i_s}$. For the non-orthogonal case, the $V^c_{i_1 i_2 \ldots i_s}$ can no longer be decomposed meaningfully into the sum of lower dimensionality terms, but would still allow a perfectly informed choice, as would the full set of the $V_{i_1 i_2 \ldots i_s}$ in the orthogonal case.

5.8   Properties of the variance based methods

In Chapter 7 we offer a description of the computational methods available in SIMLAB to estimate the variance based measures. We would like to summarise here some of the properties of these measures that correspond to the 'desired properties' of an ideal sensitivity analysis method, as described in Chapter 2.

- An ideal sensitivity analysis method should cope with the influence of scale and shape. The influence of the input should incorporate the effect of the range of input variation and the form of its probability density function (pdf). It matters whether the pdf of an input factor is uniform or normal, and what are the distribution parameters. Variance based methods meet this demand.

- A good method should allow for multidimensional averaging, contrary, for example, to what is done in computing partial derivatives, where the effect of the variation of a factor is taken when all others are kept constant at the central (nominal) value.

The measures described in this chapter instead evaluate the effect of a factor while all others are also varying, as evident from the averaging operator in their definition.

- A sensitivity measure should be model independent. The method should work regardless of the additivity or linearity of the test model. A global sensitivity measure must be able to appreciate the so-called interaction effect, especially important for non-linear, non-additive models. The property is evident with the variance based measures.

- An ideal measure should be able to treat grouped factors as if they were single factors. This property of synthesis is useful for the agility of the interpretation of the results. One would not want to be confronted with an SA made of dense tables of input–output sensitivity indices. Variance based methods are capable of grouping the factors, as we have seen already in Chapter 1. The applications below will also show the utility of such a property.

5.9 How to compute the sensitivity indices: the case of orthogonal input

In this section we offer the best recipe at our disposal to compute the full set of first-order and total-order indices for a model of k orthogonal factors. The reader who is already set on using the SIMLAB software and is not interested in numerical estimation can hence skip both this section and the next (sections 5.9 and 5.10) in this chapter. We shall not offer demonstrations of the procedure, though hints will be given to help the reader to understand how the methods work.

At first sight, the computational strategy for the estimation of conditional variances such as $V(E(Y|X_i))$ and $V(E(Y|X_i, X_j))$ would be the straightforward computation of the multidimensional integrals in the space of the input factors, writing in explicit form the E and V operators as integrals. To give an example, in computing $V(E(Y|X_i))$, the operator E would call for an integral over $\mathbf{X}_{-i}$, i.e. over all factors but X_i, including the marginal distributions for these factors, while the operator V would imply a further integral over X_i and its marginal distribution.

This approach might be needed when the input factors are non-orthogonal and it is quite expensive computationally, as customarily these integrals are estimated via Monte Carlo methods. We would need an inner Monte Carlo loop to compute, say, $E(Y|X_i = x_i^*)$, and an outer loop to apply the variance operator to obtain $V(E(Y|X_i))$. This is not necessary in the case of orthogonal factors, where the computation can be greatly accelerated. We anticipate here that we are able to obtain the full set of all k indices of the first order $\hat{S}_j$, plus all k total effect indices $\hat{S}_{Tj}$, plus each of the $\binom{k}{2}$ closed effect indices $\hat{V}^c_{-ij}$ (of order $k-2$), at the cost of $N(k+2)$ model evaluations. We have used the superscript 'hat' to denote the numeric estimates, k is the number of orthogonal factors and N is the number representative of the sample size required to compute a single estimate. To give an order of magnitude, N can vary between a few hundred to one thousand. The V^c_{-ij} closed indices are of order $k-2$, since they are closed over all factors (k) minus two (X_i, X_j), and there are $\binom{k}{2}$ of them, since this is the number of combinations of two elements from a set of k. We shall come back to this immediately below, after giving the reader a recipe for how to compute the sensitivity indices. Demonstrations can be found elsewhere (Saltelli, 2002).

Let us start with two input sample matrices $\mathbf{M}_1$ and $\mathbf{M}_2$:

$$
\mathbf{M}_1 = \begin{matrix}
x_1^{(1)} & x_2^{(1)} & \cdots & x_k^{(1)} \\
x_1^{(2)} & x_2^{(2)} & \cdots & x_k^{(2)} \\
\cdots & \cdots & \cdots & \cdots \\
x_1^{(N)} & x_2^{(N)} & \cdots & x_k^{(N)}
\end{matrix}
\tag{5.21}
$$

$$
\mathbf{M}_2 = \begin{matrix}
x_1^{(1')} & x_2^{(1')} & \cdots & x_k^{(1')} \\
x_1^{(2')} & x_2^{(2')} & \cdots & x_k^{(2')} \\
\cdots & \cdots & \cdots & \cdots \\
x_1^{(N')} & x_2^{(N')} & \cdots & x_k^{(N')}
\end{matrix}
\tag{5.22}
$$

where both matrices have row dimension k, the number of factors, and column dimension N, the sample size used for the Monte Carlo estimate. Each column of both M_1 and M_2 is a sample from the marginal distribution of the relative factor. Each row in both M_1 and M_2 is an input sample, for which a model output Y can be evaluated. All the information that we need about the form and scale of the input probability distribution functions is already contained in these matrices, which are all that we need in terms of generation of the input sample.

From M_1 and M_2 we can build a third matrix N_j:

$$
N_j =
\begin{matrix}
x_1^{(1')} & x_2^{(1')} & \cdots & x_j^{(1)} & \cdots & x_k^{(1')} \\
x_1^{(2')} & x_2^{(2')} & \cdots & x_j^{(2)} & \cdots & x_k^{(2')} \\
\cdots & \cdots & \cdots & \cdots & \cdots & \cdots \\
x_1^{(N')} & x_2^{(N')} & \cdots & x_j^{(N)} & \cdots & x_k^{(N')}
\end{matrix}
\tag{5.23}
$$

If one thinks of matrix M_1 as the 'sample' matrix, and of M_2 as the 're-sample' matrix, then N_j is the matrix where all factors except X_j are re-sampled.

$E(Y)$, the unconditional mean, and $V(Y)$, the unconditional variance, can be either estimated from values of Y computed on the sample in M_1 or in M_2, for example if M_1 is used:

$$
\hat{E}(Y) = \frac{1}{N} \sum_{r=1}^{N} f\left(x_1^{(r)}, x_2^{(r)}, \ldots, x_k^{(r)}\right)
\tag{5.24}
$$

$$
\hat{V}(Y) = \frac{1}{N-1} \sum_{r=1}^{N} f^2\left(x_1^{(r)}, x_2^{(r)}, \ldots, x_k^{(r)}\right) - \hat{E}^2(Y).
\tag{5.25}
$$

The first-order sensitivity measure for a generic factor X_j, i.e. $V(E(Y|X_j))$, can be computed as

$$
S_j = V(E(Y|X_j)) / V(Y)
\tag{5.26}
$$

$$
V(E(Y|X_j)) = U_j - E^2(Y)
\tag{5.27}
$$

where U_j can be obtained from values of Y corresponding to the sample in matrix $\mathbf{N}_j$ i.e. by:

$$\hat{U}_j = \frac{1}{N-1} \sum_{r=1}^{N} f\left(x_1^{(r)}, x_2^{(r)}, \dots, x_k^{(r)}\right)$$

$$f\left(x_1^{(r')}, x_2^{(r')}, \dots, x_{(j-1)}^{(r')}, x_j^{(r)}, x_{(j+1)}^{(r')}, \dots, x_k^{(r')}\right) \quad (5.28)$$

If again one thinks of matrix $\mathbf{M}_1$ as the 'sample' matrix, and of $\mathbf{M}_2$ as the 're-sample' matrix, then $\hat{U}_j$ is obtained from the products of values of f computed from the sample matrix *times* values of f computed from $\mathbf{N}_j$, i.e. a matrix where all factors except X_j are re-sampled. Note again that the hat is used to indicate estimates. As mentioned, we do not want to offer a demonstration of the above. Here we give a hand waiving illustration. If X_j is an influential factor, then high values of $f(x_1^{(r)}, x_2^{(r)}, \dots, x_k^{(r)})$ in Equation (5.28) above will be preferentially associated with high values of $f(x_1^{(r')}, x_2^{(r')}, \dots, x_{(j-1)}^{(r')}, x_j^{(r)}, x_{(j+1)}^{(r')}, \dots, x_k^{(r')})$. If X_j is the only influential factor (all the others being dummies) then the two values of f will be identical. If X_j is a totally non-influential factor (a dummy), then high and low values of $f(x_1^{(r)}, x_2^{(r)}, \dots, x_k^{(r)})$ will be randomly associated with high and low values of $f(x_1^{(r')}, x_2^{(r')}, \dots, x_{(j-1)}^{(r')}, x_j^{(r)}, x_{(j+1)}^{(r')}, \dots, x_k^{(r')})$. In this way, the estimate of the sensitivity measure, Equation (5.28) above, will be much higher for an influential factor X_j than for a non-influential one. It is easy to see that S_j will vary between 0 and 1, moving from a dummy (where $\hat{U}_j \approx \hat{E}^2(Y)$) to a totally influential factor (where $\hat{U}_j - \hat{E}^2(Y) \approx \hat{V}(Y)$).

Having thus discussed Equations (5.24)–(5.28), the following generalisation should be easy to grasp:

$$V_{i_1 i_2 \dots i_s}^c = V(E(Y|X_{i_1} X_{i_2} \dots X_{i_s})) = U_{i_1 i_2 \dots i_s} - E^2(Y). \quad (5.29)$$

This generalises Equations (5.26) and (5.27) to a closed index of order s. Recall that $V_{i_1 i_2 \dots i_s}^c$ is a sensitivity measure that is closed within a subset of factors, i.e. for orthogonal inputs $V_{i_1 i_2 \dots i_s}^c$ is the sum of all $V_{i_1 i_2 \dots i_s}$ terms in Equation (5.2) that is closed in the indices

$i_1, i_2, \ldots i_s$: $V_1^c = V_1$, $V_{ij}^c = V_i + V_j + V_{ij}$, and so on.

$$\hat{U}_{i_1 i_2 \ldots i_s} = \frac{1}{N-1} \sum_{r=1}^{N} f\left(x_1^{(r)}, x_2^{(r)}, \ldots, x_k^{(r)}\right)$$

$$f\left(x_{i_1}^{(r)}, x_{i_2}^{(r)}, \ldots, x_{i_s}^{(r)}, x_{l_1}^{(r')}, x_{l_2}^{(r')}, \ldots, x_{l_{k-s}}^{(r')}\right). \quad (5.30)$$

This generalises Equation (5.28). The corresponding equation for the complementary set is:

$$\hat{U}_{-i_1 i_2 \ldots i_s} = \frac{1}{N-1} \sum_{r=1}^{N} f\left(x_1^{(r)}, x_2^{(r)}, \ldots, x_k^{(r)}\right)$$

$$f\left(x_{i_1}^{(r')}, x_{i_2}^{(r')}, \ldots, x_{i_s}^{(r')}, x_{l_1}^{(r)}, x_{l_2}^{(r)}, \ldots, x_{l_{k-s}}^{(r)}\right). \quad (5.31)$$

Recall that $V^c_{-i_1 i_2 \ldots i_s}$ indicates the sum of all indices that are closed within the complementary set of $i_1, i_2, \ldots i_s$, i.e. $V^c_{-i_1 i_2 \ldots i_s} = V^c_{l_1 l_2 \ldots l_{k-s}}$ where $i_p \neq l_q$ for all $p \in [1, 2, \ldots, s]$, $q \in [1, 2, \ldots, k-s]$. Hence:

$$V^c_{l_1 l_2 \ldots l_{k-s}} = V(E(Y|X_{l_1} X_{l_2} \ldots X_{l_{k-s}})) = U_{-i_1 i_2 \ldots i_s} - E^2(Y). \quad (5.32)$$

The total effect indices can be computed from

$$\hat{S}_{Tj} = 1 - \frac{(\hat{U}_{-j} - \hat{E}^2(y))}{\hat{V}(y)} \quad (5.33)$$

where

$$\hat{U}_{-j} = \frac{1}{N-1} \sum_{r=1}^{N} f\left(x_1^{(r)}, x_2^{(r)}, \ldots, x_k^{(r)}\right)$$

$$f\left(x_1^{(r)}, x_2^{(r)}, \ldots, x_{(j-1)}^{(r)}, x_j^{(r')}, x_{(j+1)}^{(r)}, \ldots, x_k^{(r)}\right). \quad (5.34)$$

Although the above estimates might seem odd at first sight, they all obey an easy rule: all estimates are scalar products of f values. In these products, some of the factors are re-sampled (i.e. taken from $\mathbf{M}_1$ in the first f and from $\mathbf{M}_2$ in the second f) and some are not re-sampled (taken from $\mathbf{M}_1$ for both f). The rule is: do not re-sample the factors whose effect you want to estimate; for $\hat{U}_j$, needed to estimate S_j, do not resample X_j. For $\hat{U}_{-j}$, needed to estimate S_{Tj}, but based on computing the effect of non-X_j, re-sample only X_j, and so on.

Our last step in the recipe to compute the sensitivity indices is to arrange the input sample so as to keep the number of model evaluations, which we assume to be the expensive step of the analysis, to a minimum.

Let $\mathbf{a}_{i_1 i_2 \ldots i_s}$ denote the vector of length N containing model evaluations corresponding to the rows of the input factor matrix $\mathbf{N}_{i_1 i_2 \ldots i_s}$. As in Equation (5.23) above, the matrix $\mathbf{N}_{i_1 i_2 \ldots i_s}$ is obtained from matrix $\mathbf{M}_1$ by substituting all columns except $i_1, i_2, \ldots i_s$ by the corresponding columns of matrix $\mathbf{M}_2$. $\mathbf{a}_0$ will hence denote a set of model evaluations corresponding entirely to matrix $\mathbf{M}_2$, while $\mathbf{a}_{i_1 i_2 \ldots i_k}$ will indicate the vector of model evaluations corresponding entirely to matrix $\mathbf{M}_1$

$$\begin{aligned}
\mathbf{a}_0 &= f(\mathbf{M}_2) \\
\mathbf{a}_{12\ldots k} &= f(\mathbf{M}_1) \\
\mathbf{a}_{i_1 i_2 \ldots i_s} &= f(\mathbf{N}_{i_1 i_2 \ldots i_s}).
\end{aligned} \qquad (5.35)$$

Table 5.2 gives a summary of what sensitivity indices can be computed using what $\mathbf{a}_{i_1 i_2 \ldots i_s}$ vector. It corresponds to the value $k = 5$, a choice that shall be explained immediately below.

Table 5.2   Terms that can be estimated given the corresponding vectors of model evaluations, $k = 5$.

	$\mathbf{a}_0$	$\mathbf{a}_1$	$\mathbf{a}_2$	$\mathbf{a}_3$	$\mathbf{a}_4$	$\mathbf{a}_5$	$\mathbf{a}_{2345}$	$\mathbf{a}_{1345}$	$\mathbf{a}_{1245}$	$\mathbf{a}_{1235}$	$\mathbf{a}_{1234}$	$\mathbf{a}_{12345}$
$\mathbf{a}_0$	$\hat{V}$											
$\mathbf{a}_1$	$\hat{S}_{T1}$	$\hat{V}$										
$\mathbf{a}_2$	$\hat{S}_{T2}$	$\hat{V}^c_{-12}$	$\hat{V}$									
$\mathbf{a}_3$	$\hat{S}_{T3}$	$\hat{V}^c_{-13}$	$\hat{V}^c_{-23}$	$\hat{V}$								
$\mathbf{a}_4$	$\hat{S}_{T4}$	$\hat{V}^c_{-14}$	$\hat{V}^c_{-24}$	$\hat{V}^c_{-34}$	$\hat{V}$							
$\mathbf{a}_5$	$\hat{S}_{T5}$	$\hat{V}^c_{-15}$	$\hat{V}^c_{-25}$	$\hat{V}^c_{-35}$	$\hat{V}^c_{-45}$	$\hat{V}$						
$\mathbf{a}_{2345}$	$\hat{S}_1$	$\hat{E}^2$	$\hat{V}^c_{12}$	$\hat{V}^c_{13}$	$\hat{V}^c_{14}$	$\hat{V}^c_{15}$	$\hat{V}$					
$\mathbf{a}_{1345}$	$\hat{S}_2$	$\hat{V}^c_{12}$	$\hat{E}^2$	$\hat{V}^c_{23}$	$\hat{V}^c_{24}$	$\hat{V}^c_{25}$	$\hat{V}^c_{-12}$	$\hat{V}$				
$\mathbf{a}_{1245}$	$\hat{S}_3$	$\hat{V}^c_{13}$	$\hat{V}^c_{23}$	$\hat{E}^2$	$\hat{V}^c_{34}$	$\hat{V}^c_{35}$	$\hat{V}^c_{-13}$	$\hat{V}^c_{-23}$	$\hat{V}$			
$\mathbf{a}_{1235}$	$\hat{S}_4$	$\hat{V}^c_{14}$	$\hat{V}^c_{24}$	$\hat{V}^c_{34}$	$\hat{E}^2$	$\hat{V}^c_{45}$	$\hat{V}^c_{-14}$	$\hat{V}^c_{-24}$	$\hat{V}^c_{-34}$	$\hat{V}$		
$\mathbf{a}_{1234}$	$\hat{S}_5$	$\hat{V}^c_{15}$	$\hat{V}^c_{25}$	$\hat{V}^c_{35}$	$\hat{V}^c_{45}$	$\hat{E}^2$	$\hat{V}^c_{-15}$	$\hat{V}^c_{-25}$	$\hat{V}^c_{-35}$	$\hat{V}^c_{-45}$	$\hat{V}$	
$\mathbf{a}_{12345}$	$\hat{E}^2(y)$	$\hat{S}_1$	$\hat{S}_2$	$\hat{S}_3$	$\hat{S}_4$	$\hat{S}_5$	$\hat{S}_{T1}$	$\hat{S}_{T2}$	$\hat{S}_{T3}$	$\hat{S}_{T4}$	$\hat{S}_{T5}$	$\hat{V}$

A few notes:

1. Table 5.2 can be interpreted by referring to the equations just given in this section. For example, we have labelled the entry corresponding to the intersection $\mathbf{a}_0$ and $\mathbf{a}_1$ as $\hat{S}_{T1}$, as $\mathbf{a}_0 \cdot \mathbf{a}_1$ yields $\hat{U}_{-1}$ that in turn can be used to compute $\hat{S}_{T1}$ (Equation (5.34)) and so on for the other terms.

2. The diagonal has been labelled as providing an estimate of $\hat{V}(Y)$, as this is what can be obtained by the scalar product $\mathbf{a}^2_{i_1i_2...i_s}$. In fact each of the $2k + 2$ vectors $\mathbf{a}_{i_1i_2...i_s}$ can yield an estimate of $\hat{E}(Y)$. The known $\hat{E}(Y)$ for each $\mathbf{a}_{i_1i_2...i_s}$ can again be used to estimate $\hat{V}(Y)$.

3. The intersection of vectors $\mathbf{a}_1$ and $\mathbf{a}_{2345}$ has been labelled as an estimate of $\hat{E}^2(Y)$, as all columns in the two sampling matrices are different and the scalar product $\mathbf{a}_{i_1i_2...i_s}\mathbf{a}_{j_1j_2...j_r}$ provides an estimate of the square of $E(Y)$. We shall come back to this point in a moment.

4. The two vectors $\mathbf{a}_2$ and $\mathbf{a}_{2345}$ allow the computation of $\hat{V}^c_{12}$ as columns 1 and 2 are identical in the two sampling matrices.

5. The two vectors $\mathbf{a}_{2345}$ and $\mathbf{a}_{1345}$ allow the computation of $V^c_{345} = \hat{V}^c_{-12}$ as columns 3, 4 and 5 are identical in the two sampling matrices.

The shaded cells in the table are those whose computation is suggested, i.e. our recommendation to the reader is to evaluate the model output Y for the seven vectors $\{\mathbf{a}_0, \mathbf{a}_1, \mathbf{a}_2, \mathbf{a}_3, \mathbf{a}_4, \mathbf{a}_5, \mathbf{a}_{12345}\}$. It is easy to see that this allows the computation of all the $\hat{S}_{Tj}$s and $\hat{S}_j$ indices, with $j \in [1, 2, 3, 4, 5]$ at the cost of $N(k + 2) = 7N$ model evaluations. Furthermore, in this way we have produced one estimate for each of the $\binom{5}{3} = 10$ indices, $\hat{V}^c_{-ij}$, complementary to the second-order ones, that for $k = 5$ happen to be closed indices of the third order. Note that for $k = 6$ we would have obtained one estimate for each of the $\binom{6}{4} = 15$ closed indices of the fourth order and so on for larger values of k, based on the known property that $\binom{k}{j} = \binom{k}{k-j}$ for $k \geq j$.

The reader may easily discover arrangements alternative to that suggested in the grey cells. The table also shows that if one were to decide to compute all vectors in the table at a cost of $(2k + 2)N$ model runs, i.e. not only $\{a_0, a_1, a_2, a_3, a_4, a_5, a_{12345}\}$ but also $\{a_{2345}, a_{1345}, a_{1245}, a_{1235}, a_{1234}\}$, one would obtain double, rather than single, estimates for each of the $\hat{S}_{Tj}$s, $\hat{S}_j$ and $\hat{V}_{-ij}^c$. Additionally one would obtain double estimates for each of the $\binom{5}{2} = 10$ closed indices of the second order. The reader may easily verify as an exercise that for $k = 4$, Table 5.2 would look somehow different: we would obtain more estimates of the second-order terms. For $k = 3$ we would obtain more estimates of the first-order terms. In conclusion, the illustration offered with $k = 5$ is of more general use.

For the reader who wants to re-code the algorithms in this section, we offer without demonstration an additional tip. This is to compute two different estimates of the squared mean, which is needed to compute all indices:

$$\hat{E}^2 = \left(\frac{1}{N} \sum_{r=1}^{n} f\left(x_1^{(r)}, x_2^{(r)}, \ldots, x_k^{(r)}\right) \right)^2 \qquad (5.36)$$

and

$$\hat{E}^2 = \frac{1}{N} \sum_{r=1}^{n} f\left(x_1^{(r)}, x_2^{(r)}, \ldots, x_k^{(r)}\right) f\left(x_1^{(r')}, x_2^{(r')}, \ldots, x_k^{(r')}\right). \qquad (5.37)$$

Equation (5.36) is simply Equation (5.24) squared. Equation (5.37) is also a valid estimate, based on the product $a_0 \cdot a_{12...k}$, i.e. on matrices M_1 and M_2. Equation (5.37) should be used in Equation (5.25), i.e. to compute the estimates of first-order indices $\hat{S}_j$, while Equation (5.36) should be used in Equation (5.33), to estimate the total indices $\hat{S}_{Tj}$.

We have given no prescription about how to build the sample matrices M_1 and M_2, which form the basis for the analysis. As mentioned, each column in these matrices represents an independent sample from the marginal distribution of a factor. Usually these can be obtained via standard statistical packages that provide, given a set of random numbers $\varsigma^{(r)} \in [0, 1]$, $r = 1, 2, \ldots N$, the corresponding factor values $x_j^{(r)}$ as a solution of $\varsigma^{(r)} = \int_{-\infty}^{x_j^{(r)}} p(x_j) dx_j$, where $p(x_j)$ is the marginal probability density function of factor X_j.

5.9.1   *A digression on the Fourier Amplitude Sensitivity Test (FAST)*

A very elegant estimation procedure for the first-order indices, S_i, and orthogonal input is the classical FAST, developed by Cukier *et al.* (1973) and then systematised by Cukier *et al.* (1978) and Koda *et al.* (1979).

An extension to FAST was proposed by Saltelli *et al.* (1999) for the estimation of the total sensitivity index, S_{Ti}. The advantage of the extended FAST is that it allows the simultaneous computation of the first and total effect indices for a given factor X_i. We refer to the papers just cited for a thorough explanation of the techniques. Both the classical FAST and the extended FAST are implemented in SIMLAB.

5.10   How to compute the sensitivity indices: the case of non-orthogonal input

When the uncertain factors X_i are non-orthogonal, the output variance cannot be decomposed as in Equation (5.2). Consequently, the computational shortcuts available for orthogonal inputs, described in the previous section, are no longer applicable. $V_i = V(E(Y|X_i))$ (or $S_i = V_i/V(Y)$) is still the sensitivity measure for X_i in terms of Setting FP, but now V_i also carries over the effects of other factors that can, for example, be positively or negatively correlated to X_i. So, the sum of all the V_i across all the inputs might be higher than the unconditional variance $V(Y)$, as already discussed. The term $V(E(Y|X_i, X_j))$ (or V_{ij}^c for brevity) is the measure of the joint effect of the pair (X_i, X_j). However, for non-orthogonal input we cannot write that $V_{ij}^c = V_i + V_j + V_{ij}$, i.e. relate V_{ij}^c to the sum of the individual effects V_i and V_j, and a pure interaction term. Higher order terms (e.g., $V_{ijlm...}^c$) have similar meaning. A number of approaches exist in order to estimate the partial variances V_i, V_{ij}^c, V_{ijl}^c, etc., as discussed next.

The terms V_i, V_{ij}^c, V_{ijl}^c, V_{ijlm}^c, etc. can be estimated by means of multidimensional integrals in the space of the input factors, writing in explicit form the operators E and V (brute force approach). The integrals can be estimated via Monte Carlo methods. However, this approach is quite computationally expensive. To estimate $V_i = V(E(Y|X_i))$ for a given input factor X_i we have to fix X_i at a value x_i^* selected within its range of uncertainty, and evaluate the $(k-1)$-dimensional integral $E(Y|X_i = x_i^*)$ by Monte Carlo sampling over the space $(X_1, \ldots, X_{i-1}, x_i^*, X_{i+1}, \ldots X_k)$. This step may require, say, $N = 100$ sample points (which corresponds to N model evaluations). Then, this procedure has to be repeated by fixing X_i at different values (say again $r = 100$ to give an idea), to explore the X_i axis. This means that r different integrals of dimension $k-1$ have to be estimated, and it implies that the total number of model runs is Nr.

The final step is to estimate the variance of the r conditional expectations $E(Y|X_i = x_i^*)$. This last integral is just a one-dimensional integral over X_i and does not require any new model run. Given that this has to be performed for each factor in turn, we would need to make kNr runs of the model.

The same procedure applies for the terms V_{ij}^c. In this case we fix the pair (X_i, X_j) at a value (x_i^*, x_j^*), and evaluate the $(k-2)$-dimensional integral $E(Y|X_i = x_i^*, X_j = x_j^*)$ by Monte Carlo sampling over the space $(X_1, \ldots, X_{i-1}, x_i^*, X_{i+1}, \ldots, X_{j-1}, x_j^*, X_{j+1}, \ldots X_k)$. This step again requires N sample points (i.e., model runs), and the procedure is repeated by considering r different pairs (x_i^*, x_j^*). The variance of the r conditional expectations $E(Y|X_i = x_i^*, X_j = x_j^*)$ is a two-dimensional integral over (X_i, X_j) and does not require any further model run. There are $\frac{1}{2}k(k-1)$ possible combinations of second-order terms V_{ij}^c. If we were interested in all of them, we would need to execute our model $\frac{1}{2}k(k-1)Nr$ times.

A problem that we have not tackled here is how to generate samples from conditional distributions. For example, to estimate V_i we need to generate a sample from the space $(X_1, \ldots, X_{i-1}, x_i^*, X_{i+1}, \ldots X_k)$. Usually, some form of rejection sampling method is used, such as Markov Chain Monte Carlo.

However, this is not the approach implemented in SIMLAB and is hence not described here.

A rather efficient estimation procedure is available for the first-order terms that is due to McKay (1995) and uses the replicated Latin hypercube sampling design (r-LHS, McKay *et al.*, 1979). The procedure that follows, implemented in SIMLAB, is an extension of McKay's approach to the non-orthogonal case (Ratto and Tarantola 2003). Note that in McKay's approach the cost of the estimate is not kNr but simply Nr, as the same sample can be used to estimate main effects for all factors, as described below.

When using simple LHS, the range of variability of each factor is divided into N non-overlapping intervals, or bins, of equal probability, i.e. such that the probability for that factor to fall in any of the bins is exactly $1/N$. Imagine we generate, independently one from another, r such samples (replicated LHS, or r-LHS). Consider a factor X_j and let $p_i(Y|X_j = x^*_{j,i})$ be the conditional distribution of Y when X_j is fixed to the value $x^*_{j,i}$ in the ith bin. Let $(y_j^{(li)}, l = 1, 2, \ldots r)$ be a value of Y corresponding to replica l from that distribution. There will be in total rN such points in the r-LHS design, as $i = 1, 2, \ldots N$ (the bins) and $l = 1, 2, \ldots r$ (the replicas), see Figure 5.2.

The sample conditional and unconditional means are:

$$E(Y|X_j = x^*_{j,i}) = \bar{y}_j^{(i)} = \frac{1}{r} \sum_{l=1}^{r} y_j^{(li)} \qquad (5.38)$$

$$E(Y) = \bar{y} = \frac{1}{N} \sum_{i=1}^{N} \bar{y}_j^{(i)}. \qquad (5.39)$$

The main effect sensitivity index is computed as:

$$S_j = \frac{V_j}{V} \qquad (5.40)$$

$$V_j = \frac{1}{N} \sum_{i=1}^{N} (\bar{y}_j^{(i)} - \bar{y})^2 \qquad (5.41)$$

$$V = \frac{1}{Nr} \sum_{i=1}^{N} \sum_{l=1}^{r} (y_j^{(li)} - \bar{y})^2. \qquad (5.42)$$

The SIMLAB routines that implement this procedure also include a correction term to reduce bias, due to McKay (1995).

One feature of a non-orthogonal r-LHS sample is that in drawing r different samples of dimension N, we take care of selecting for each factor always the *same cell* in each bin. This way, all r LHS samples use the same values for each factor in each bin (see Figure 5.2). For orthogonal inputs this could be done simply by making r independent random permutations of the columns of a base LHS sample, while for non-orthogonal inputs it is necessary to generate r different LHS samples, in order to preserve the dependency structure that would be destroyed by the random permutation. We can take the inner conditional expectations $E(Y|X_i = x_i^*)$ in *each* bin and for *any* factor, X_j, using a *unique* r-LHS design to compute all the indices, S_j. The total number of runs is therefore as high as rN for the r-LHS design, reducing the cost with respect to a brute force approach by a factor k. The methods for generating non-orthogonal LHS samples are the method of Iman and Conover (1982), for creating LHS samples with a user-defined rank correlation structure, and the method of Stein (1987), for creating LHS samples from any type of non-orthogonal sample, both implemented in SIMLAB.

When generating the sample it is suggested that one takes N always larger than r; it is also advisable to use values of r that are not greater than 50 and use N as the reference parameter to increase asymptotic convergence.

Another approach, not implemented in SIMLAB yet, is to apply the estimator in Equation (5.28) for orthogonal input to the non-orthogonal case, using the r-LHS design. This can be found in Ratto and Tarantola (2003). It is evident from this section that a re-coding of the algorithms for sensitivity indices for the non-orthogonal case is not straightforward. For example, one has to code LHS and the Iman and Conover procedure for generating rank-correlated input samples. Public domain software (in FORTRAN) for the latter is available (Iman and Shortencarier 1984). In general, we would recommend the reader to recur to non-orthogonal input only when essential, as convergence of sensitivity estimates is much slower for the non-orthogonal cases than for the orthogonal ones, and bias has to be taken care of.

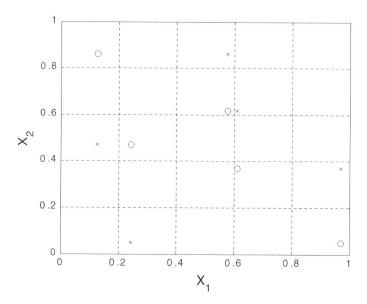

Figure 5.2   Example of r-LHS sampling for two variables the number of bins $N = 5$, the number of replicates $r = 2$. In each replicate, the same values for each factor are used in each bin.

5.11   Putting the method to work: the Level E model

Before we apply the variance based sensitivity measures to a test case, we would like to recall briefly some of their properties that will come handy in the interpretation of the results.

1. Whatever the correlations/dependencies among the factors and the interactions in the model, S_i gives how much one could reduce the variance on average if one could fix X_i.

2. Whatever the correlations/dependencies among the factors and the interactions in the model, $S^c_{i_1 i_2 ... i_s}$ gives how much one could reduce the variance on average if one could fix $X_{i_1}, X_{i_2}, \ldots X_{i_s}$.

3. With orthogonal input, S_{Ti} is greater than S_i (or equal to S_i in the case where X_i is not involved in any interaction with other input factors). The difference $S_{Ti} - S_i$ is a measure of how much X_i is involved in interactions with any other input variable. In the presence of non-orthogonal input, S_{Ti} can be lower than S_i.

4. Whatever the dependency structure among the factors and the interactions in the model, $S_{Ti} = 0$ implies that X_i is non-influential.

5. With orthogonal input, the sum of all the S_is is less than 1 for non-additive models and equal to 1 for additive ones. The difference $1 - \sum_i S_i$ gives a measure of the interactions of any order. This is not the case for non-orthogonal inputs.

In all the tests in this section, which concern the Level E model described in Chapter 3, the output of interest is the total radiological dose $Y(t)$ from Equation (3.15). The radiological dose is predicted at 26 time points in the range $(10^3\text{--}10^7 \text{ yr})$. The model includes twelve uncertain input variables, which are listed in Table 3.5, which we reproduce below as Table 5.3.

We want to achieve two objectives in this exercise:

- To identify non-relevant input factors for model reduction purposes (i.e. Setting FF). This implies calculating total effect, S_{Ti} sensitivity indices for individual input factors. If $S_{Ti} \approx 0$ for all the time points, the input factor X_i does not influence the model output at any time point. Therefore the factor X_i can be frozen to any value within its range of variation because it does not contribute to the output variance neither singularly or in combination with other input factors.

- To identify relevant input factors for subsequent calibration / optimisation tasks, or for prioritisation of research (i.e. Setting FP). To achieve this objective we need to estimate the $S_i \quad \forall i = 1, \ldots, 12$ for all the input factors. A high value for the S_i indicates an input factor X_i that drives the model output variance consistently. This can be seen as indicating where to direct effort in the future to reduce that uncertainty.

5.11.1 Case of orthogonal input factors

Let us first focus on the case of orthogonal input factors, as the approach to the calculation of the sensitivity indices is much easier to tackle.

Table 5.3   List of input factors for the Level E Model

Notation	Definition	Distribution	Range	Units
T	Containment time	Uniform	$[100, 1000]$	yr
k_I	Leach rate for iodine	Log-Uniform	$[10^{-3}, 10^{-2}]$	mol/yr
k_C	Leach rate for Np chain nuclides	Log-Uniform	$[10^{-6}, 10^{-5}]$	mol/yr
$v^{(1)}$	Water velocity in the first geosphere layer	Log-Uniform	$[10^{-3}, 10^{-1}]$	m/yr
$l^{(1)}$	Length of the first geosphere layer	Uniform	$[100, 500]$	m
$R_I^{(1)}$	Retention factor for iodine in the first layer	Uniform	$[1, 5]$	—
$R_C^{(1)}$	Retention factor for the chain elements in the first layer	Uniform	$[3, 30]$	—
$v^{(2)}$	Water velocity in the second geosphere layer	Log-Uniform	$[10^{-2}, 10^{-1}]$	m/yr
$l^{(2)}$	Length of the second geosphere layer	Uniform	$[50, 200]$	m
$R_I^{(2)}$	Retention factor for iodine in the second layer	Uniform	$[1, 5]$	—
$R_C^{(2)}$	Retention factor for the chain elements in the second layer	Uniform	$[3, 30]$	—
W	Stream flow rate	Log-Uniform	$[10^5, 10^7]$	m^3/yr
C_I^0	Initial inventory for ^{129}I	Constant	100	mol
C_{Np}^0	Initial inventory for ^{237}Np	Constant	1000	mol
C_U^0	Initial inventory for ^{233}U	Constant	100	mol
C_{Th}^0	Initial inventory for ^{229}Th	Constant	1000	mol
w	Water ingestion rate	Constant	0.73	m^3/yr
β_I	Ingestion-dose factor for ^{129}I	Constant	56	Sv/mol
β_{Np}	Ingestion-dose factor for ^{237}Np	Constant	6.8×10^3	Sv/mol
β_U	Ingestion-dose factor for ^{233}U	Constant	5.9×10^3	Sv/mol
β_{Th}	Ingestion-dose factor for ^{229}Th	Constant	1.8×10^6	Sv/mol

---

Box 5.1   Level E

The user might want to run a small-sample Monte Carlo analysis to estimate preliminary model output statistics (including shape of model output distribution). This Monte Carlo analysis is shown in this Box. Let us generate a random sample of size N = 1024 over the space of 12 input factors and run the

Level E model. The model output mean and 90% uncertainty bounds are given in the figure below:

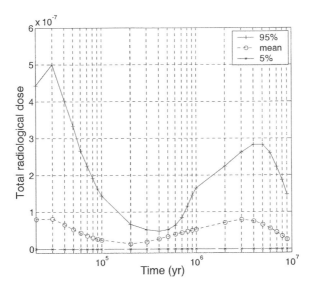

The figure shows the dynamics of the total output dose, characterised by two maxima, corresponding to the release of two different isotopes ^{129}I (fast dynamics) and ^{239}Np (slower dynamics) respectively. The second maximum has smaller peak values, but in terms of mean path it is comparable to the first one.

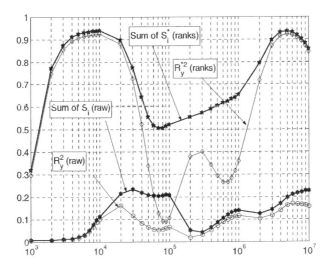

With these model runs we can also compute the Standardised Regression Coefficients (SRC's), the Standardised Rank Regression Coefficients (SRRC's) and the corresponding coefficients of determination across time.[3] The results are shown in the second figure above. The figure indicates that the model has a strong non-additive, non-monotonic behaviour, especially around 10^5 yr. The effect of studying ranks instead of raw values is such that all monotonic input–output relationships are linearised (see also Saltelli and Sobol' (1995)). This means that, for example, an exponential relationship, which would be badly identified by SRCs, would provide a very high SRRC. So we can say that the R_y^2 of SRCs measures the degree of linearity of the model, while the R_y^{*2} of SRRCs measures the degree of *monotonicity* of the additive components.

In the same figure, we also show the cumulative sums of the first-order sensitivity indices, for both the raw values, S_i, and the ranks, S_i^*. Keeping in mind that the sum of the first-order sensitivity indices measures the degree of additivity of the model, regardless of the monotonicity of the input–output relationship, it is interesting to note that:

1. the values for ranks are always larger than the values for raw values;

2. as expected, the cumulative sum of the first-order indices is always larger than the corresponding R_y^2;

3. for raw values, when the difference between the cumulative sum of the S_is and R_y^2 is small, the additive components of the model are mainly linear;

4. for ranks, when the difference between the cumulative sum of the S_i^*s and R_y^{*2} is small, the additive components of the model are mainly monotonic;

5. for raw values, when the cumulative sum of the S_is is large and R_y^2 is small, the model has a significant additive

---

[3] SRRC's are the same as SRC's but on ranks instead than on raw values.

component which is either non-linear or non-monotonic, implying that SRCs are ineffective;

6. for ranks, when the cumulative sum of the S_i^*s is large and R_y^{*2}s is small, the model has a significant additive component which is non-monotonic, implying that SRRCs are ineffective;

7. when the cumulative sum of the S_i^*s is large (yet always <1!) and the cumulative sum of the S_is is small (as in the Level E model), this is usually due to outputs characterised by distributions with very long tails, whose extreme values are connected to particular combinations of factors, i.e. the non-additive components of the model (interactions) have peaks concentrated in small regions of the input factor space, where the output varies by orders of magnitude. This extreme effect of interactions can be drastically smoothed using ranks;

8. when the model is additive, *both* the cumulative sums of the S_i^*s and S_is must be equal to 1, apart from numerical error.

We use a base sample of $N = 550$ points. At the total cost of $N \times (2k + 2) = 550 \times 26 = 14\,300$ model evaluations, the method allows us to estimate all the S_i and the S_{Ti} (shown in Table 5.4 for the most important factors) and, therefore, to fulfil both the objectives set out before.

First objective
For time dependent model outputs it is custom and practice to display the sensitivity indices as cumulative area plots. The area plot for the total indices is shown in Figure 5.3. We can see that the parameters k_C, T and k_I are non-influential at any time point, i.e. their corresponding areas are just lines. Therefore they can be fixed in a subsequent analysis and the dimensionality of the space of the input factors can be reduced from 12 to 9. The variables $v^{(1)}$, $l^{(1)}$, W, $R_C^{(1)}$ and $R_I^{(1)}$ show high values for their total indices along almost all the time range. This means that an input factor might influence the output through interactions with other input factors.

Table 5.4   First-order and total effect sensitivity indices obtained with $N = 550$ for the three most important variables.

Time (yr)	$S_{V(1)}$	$S_{TV(1)}$	$S_{TV(1)} - S_{V(1)}$	$S_{L(1)}$	$S_{TL(1)}$	$S_{TL(1)} - S_{L(1)}$	S_W	S_{TW}	$S_{TW} - S_W$
20 000	0.1211	0.8044	0.6833	0.0065	0.4324	0.4259	0.1027	0.5659	0.4632
40 000	0.0736	0.8301	0.7565	−0.0199	0.3913	0.4112	0.2077	0.6448	0.4371
60 000	0.0024	0.848	0.8456	−0.0063	0.3808	0.3871	0.1117	0.5859	0.4742
80 000	0.0158	0.9389	0.9231	0.0098	0.4657	0.4559	0.1368	0.579	0.4422
100 000	0.1649	0.8967	0.7318	0.0041	0.4893	0.4852	0.0983	0.5379	0.4396
200 000	0.0256	0.9649	0.9393	−0.0013	0.6492	0.6505	0.1249	0.476	0.3511
400 000	0.0672	0.8863	0.8191	0.0055	0.5002	0.4947	0.0126	0.5058	0.4932
600 000	0.0553	0.7171	0.6618	−0.0027	0.3926	0.3953	0.0014	0.5068	0.5054
800 000	0.0568	0.6629	0.6061	0.006	0.3713	0.3653	0.0026	0.4988	0.4962
1 000 000	0.0618	0.6541	0.5923	0.0154	0.3558	0.3404	0.0039	0.4893	0.4854
2 000 000	0.0729	0.6729	0.6	0.0163	0.3719	0.3556	0.0043	0.4755	0.4712
4 000 000	0.047	0.6105	0.5635	0.0225	0.3114	0.2889	0.0079	0.5351	0.5272
6 000 000	0.0423	0.6857	0.6434	0.0157	0.4374	0.4217	0.0555	0.6022	0.5467
8 000 000	0.087	0.68	0.593	−0.0142	0.3277	0.3419	0.1108	0.6937	0.5829

Negative signs in the table are due to numerical errors in the Sobol' estimates. Such negative values can often be encountered for the Sobol' method when the analytical sensitivity indices are close to zero (i.e. for unimportant factors). Increasing the sample size of the analysis reduces the probability of having negative estimates. FAST estimates are always positive, by construction.

Second objective

We display the first order indices, S_i, in Figure 5.4. The input factors with high first-order effects are $v^{(1)}$, W and $l^{(1)}$. A high value for S_i corresponds to an input factor X_i, giving a consistent contribution to the model output variance. S_i can be high in some time intervals and low in others. In the example, the input factors $v^{(1)}$ and W jointly account for about 20% of the output variance up to time $= 100\,000$ years. Then, at 200 000 years, there is a strong decrease in the output variance explained by the first-order effects and, correspondingly, an increase in the total effects. The sum of the first-order indices is less then 5% of the overall output variance: the output seems to be driven mostly by interactions between the factors. Around 200 000 years, the model coefficient of determination on ranks is also very low. All these are symptoms that something is happening in the model. Actually, we are in a transition phase where the fast dynamics is ending and the slow

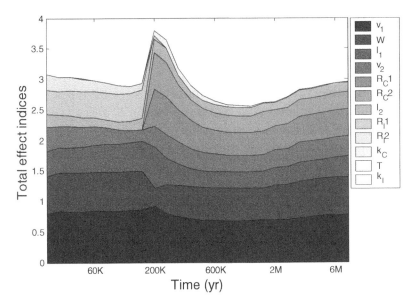

Figure 5.3 Area plot of asymptotic values for the total sensitivity indices (obtained with $N \gg 1000$) for the Level E model. The areas are cumulated from the most important factors (areas at the bottom) to the least ones (areas at the top). In the legend, factors are ordered by importance from the top (important) to the bottom (unimportant).

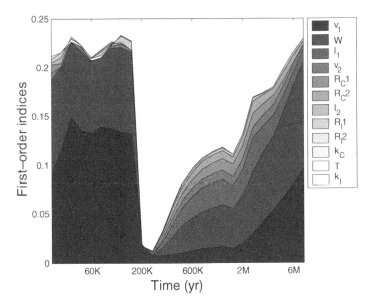

Figure 5.4 Area plot of asymptotic values for the first-order sensitivity indices (obtained with $N \gg 1\,000$) for the Level E model.

dynamics is starting. After that, the first-order indices return at higher values.

We can also investigate whether the input factor X_i is involved in interactions with other input factors. The difference $S_{Ti} - S_i$, also shown in Table 5.3 for the most important variables, is a measure of the pure interaction between a given factor X_i and all the others. Highlighting interactions among variables helps to improve our understanding of the model structure. For example, the factor $v^{(1)}$ is important through the interactions with other factors. We will discover other interesting features of $v^{(1)}$ in Chapter 6.

5.11.2 Case of correlated input factors

Let us now consider the Level E with the input correlation structure given in Table 3.6, repeated below as Table 5.5.

We use the estimator described in Section 5.10 to estimate all the S_i. For the r-LHS design we select a base sample of size $N = 1000$ with r = 20 replicates.

The total cost of estimating all the S_i is hence $N \times r = 20\,000$.

The results of the uncertainty analysis are quite similar to that for the orthogonal case. The results of the sensitivity analysis are shown in Figure 5.5 for the time range t = (20 000–9 000 000) years. While, in the orthogonal case, the factor $v^{(1)}$ was the most important one in terms of S_i, now the high correlation with both T and $v^{(2)}$, which are less influential, makes $v^{(1)}$ less influential too. In

Table 5.5   Configuration for correlated input of the Level E model.

Pairs of correlated factors	Rank correlation
k_I, k_C	0.5
$R_I^{(1)}, R_C^{(1)}$	0.3
$R_I^{(2)}, R_C^{(2)}$	0.3
$T, v^{(1)}$	−.7
$v^{(1)}, v^{(2)}$	0.5
$R_I^{(1)}, R_I^{(2)}$	0.5
$R_C^{(1)}, R_C^{(2)}$	0.5

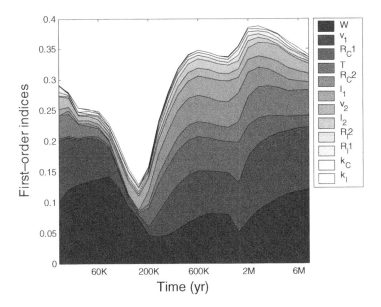

Figure 5.5 Area plot of the first-order sensitivity indices, S_i, for the Level E model and correlated input at $N \times r = 20\,000$.

fact, W has the largest S_i all through the dynamic of iodine. After $t = 200\,000$ years, $v^{(1)}$ and W have similar values of S_i. Unlike the orthogonal case where $v^{(1)}$ and W account for most of the output uncertainty, some influence of the other factors can be noted in Figure 5.5. These factors contribute to the output uncertainty thanks to the correlations in which they are involved.

The omission of correlations in the original specification of the Level E exercise was legitimate as far as uncertainty analysis is concerned, as the output dose does not change much, but there was an over-simplification as far as the identification of the influential factors is concerned.

5.12 Putting the method to work: the bungee jumping model

We consider here the simple model bungee jumping, presented in Chapter 3. We aim at minimising the variable h_{min} (i.e. the minimum distance to the asphalt during the oscillation) given the uncertainty in the three factors H, M and σ. The lower h_{min} is, the higher will be the risk (and the excitement) of the jump. We apply sensi-

Table 5.6   List of input factors for the bungee jumping model.

Notation	Definition	Distribution	Range	Unit
H	The distance from the platform to the asphalt	Uniform	[40; 60]	m
M	Our mass	Uniform	[67; 74]	kg
σ	The number of strands in the cord	Uniform	[20; 40]	–

tivity analysis to see how input factors influence the uncertainty of $h_{\min}$ in terms of setting FP.

The initial configuration for the input factors is given in Table 5.6

We remind ourselves that the model has the simple form given in Equation (3.1):

$$h_{\min} = H - \frac{2Mg}{k_{el}\sigma}.$$

We run a sensitivity analysis with the classic FAST method. We select 1000 sample points (which coincides with the total cost of the analysis in the case of FAST). The method provides the first order indices, S_i, (see Table 5.7) and also the uncertainty analysis (see Figure 5.6) that show that the jump is not always successful.

We also test the method of Sobol' on the same example. We use a base sample of $N = 256$ points. At the total cost of $N \times (k + 2) = 256 \times 5 = 1280$ model runs, the method provides all S_i and S_{Ti} (see again Table 5.7), though, in setting FP, we are interested only in the S_i.

The estimates of S_i are very close for both methods. For orthogonal inputs, the total indices cannot be, by definition, smaller than the corresponding first-order indices. However, due to numerical error, this can occur for the Sobol' estimates at a small base sample size.

Let us take into account the results of FAST. We can see that the most important factor is σ, which approximately accounts for 55% of the variance of $h_{\min}$. This means that, in order to achieve our objective (i.e., to reduce the uncertainty of $h_{\min}$ below certain levels), we should try to reduce the uncertainty of σ.

The distance platform–asphalt, H, is important and we should try to get better knowledge for this variable as well. However, we

Table 5.7 Estimates of sensitivity indices for the bungee-jumping test case.

Input factor	S_i(FAST)	S_i(Sobol')	S_{Ti}(Sobol')
H	0.4370	0.4511	0.4538
M	0.0114	0.0116	0.0115
σ	0.5458	0.5610	.5595

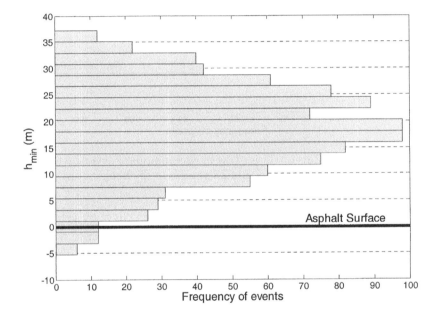

Figure 5.6 Uncertainty analysis of the bungee jumping excitement indicator h_{min} with three uncertain factors.

would be wasting time in trying to improve the knowledge of our mass, as its effect on the uncertainty of h_{min} is negligible.

Adding up the three sensitivity indices, we can see that their sum is very close to 1 (the method of Sobol' yields a number that is greater than one: this is again due to the small sample size used to estimate the indices). This means that, in spite of its analytic formulation, the model is almost fully additive, i.e. no interaction emerges between the variables M and σ. As an exercise for the reader, we suggest finding when the interactions emerge by changing the input distributions.

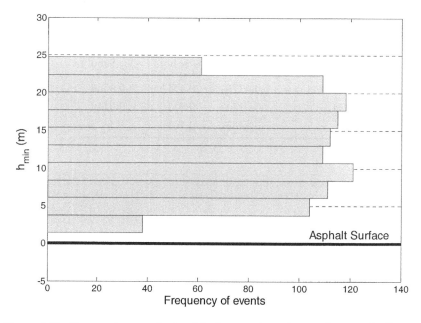

Figure 5.7 Uncertainty analysis of the bungee jumping excitement indicator $h_{\min}$ with two uncertain factors and 25 strands in the bungee cord.

What happens if we remove the uncertainty on σ by selecting a specific cord? Let us select $\sigma = 25$ strands and execute a Monte Carlo uncertainty analysis by generating 1000 random points over the space (H, M) and evaluating the model. We get a safe jump $(SJ = 100\%)$ with $h_{\min}$ ranging between 1.3 m and 24.7 m above the asphalt. Figure 5.7 gives the distribution of $h_{\min}$ after the number of strands has been fixed at 25. The standard deviation of $h_{\min}$ has decreased from 8.72 m (in the previous situation, see Figure 5.6) to 6.1 m. The jump seems to be more exciting now.

5.13 Caveats

The variance-based methods are extremely powerful in quantifying the relative importance of input factors or of groups of them. The main drawback is the cost of the analysis, which, in the case of computationally intensive models, can become prohibitive. In such situations we suggest using the revised Morris method, which, as discussed in Chapter 4, provides a quick response in terms of the measures μ^*s that are best parallel to total sensitivity indices. This allows fixing a subset of model input factors and, if the model is not too expensive, carrying out, on the reduced set of factors, a more parsimonious analysis with variance based methods. In the case of extremely computationally expensive models, the revised Morris method alone can provide a satisfactory sensitivity portrait.

We have mainly concentrated on applications and methods of computation for orthogonal inputs because they are more efficient and also because the introduction of a dependency structure in the input factors makes the interpretation of results more difficult, as discussed in Chapter 1. For example, a strong interaction in the model can be compensated for by a correlation structure in such a way that first-order and total effects become equal, leading one to a misleading conclusion that the model is additive. For this reason we suggest that one always performs a first analysis with orthogonal inputs. The analysis of the effects of non-orthogonality in the input factors should be postponed at a second stage, only if deemed necessary. In this way the interpretation of results would be easier and unambiguous.

6 SENSITIVITY ANALYSIS IN DIAGNOSTIC MODELLING: MONTE CARLO FILTERING AND REGIONALISED SENSITIVITY ANALYSIS, BAYESIAN UNCERTAINTY ESTIMATION AND GLOBAL SENSITIVITY ANALYSIS

In the preceding chapters we have mainly focused on a prognostic use of models. In this chapter we would like to return to a diagnostic use of models, i.e. in model calibration, with the purpose of describing in some detail the sensitivity issues pertaining to it. We start by showing how this is related to the Factors Mapping Setting, introduced in Chapter 2. Then two classes of methods, Monte Carlo filtering and Bayesian estimation, are discussed, showing how sensitivity analysis can tackle most of the critical questions raised by the calibration exercise.

6.1 Model calibration and Factors Mapping Setting

As some readers may have noted, there is some resonance between the quantitative sensitivity approaches presented in this book and the Monte Carlo-based approaches to model calibration that have appeared in the literature in the last two decades, mainly in the framework of environmental sciences. We refer here to the problem of an analyst trying to 'adjust' his model to some 'acceptable behaviour'. Even if classical model fitting falls under this kind of

Sensitivity Analysis in Practice: A Guide to Assessing Scientific Models A. Saltelli, S. Tarantola, F. Campolongo and M. Ratto © 2004 John Wiley & Sons, Ltd. ISBN 0-470-87093-1

problem, we address here the broader contexts of model calibration and acceptability, by allowing for, e.g., the use of qualitative definitions expressed in terms of thresholds, based on 'theoretical' (physical, chemical, biological, economical, etc.) constraints, expert opinions, legislation, etc.

Mechanistic models used in many scientific contexts (e.g. environmental sciences), based on traditional scientific descriptions of component processes, almost always contain ill-defined parameters and are thus referred to as over-parameterised models (e.g. Draper and Smith, 1981, p. 487). Accordingly, it is often concluded that the estimation of a unique set of parameters, optimising goodness of fit criteria given the observations, is not possible. Moreover, different competing model structures (different constitutive equations, different types of process considered, spatial/temporal resolution, etc.) are generally available that are compatible with the same empirical evidence (see e.g. Hornberger and Spear, 1981).

This implies the unfeasibility of the traditional estimation/fitting approach. The investigator is then referred to the 'weaker' categorisation into acceptable/ unacceptable behaviour. As a result one needs to establish *magnitude* and *sources* of model prediction uncertainty, as well as the characterisation of model acceptable properties, i.e. which assumptions, structures or combinations of model parameters are compatible with the defined acceptability criteria. This is done with Monte Carlo simulation analyses, which can be divided into two big classes: Monte Carlo filtering and Bayesian analysis. Both approaches entail an uncertainty analysis followed by a sensitivity analysis, which now assumes a peculiar and critical value. In fact, the scope of sensitivity analysis is not only to quantify and rank in order of importance the sources of prediction uncertainty, similarly to the settings discussed in the previous chapters, but, what is much more relevant to calibration, to identify the elements (parameters, assumptions, structures, etc.) that are mostly responsible for the model realisations in the acceptable range.

Where a classical estimation approach is impractical and model factors cannot be defined, for example, by an estimate minus/plus some standard error, or the clear definition of *a well defined* model structure or set of hypotheses cannot be established, applying, e.g., standard statistical testing procedures, sensitivity analysis becomes an essential tool. Model factors can be classified, for example, as 'important/unimportant' according to their capability of driving the model behaviour. Such capability is clearly highlighted by the sensitivity analysis, which plays a similar role, e.g., of a t-test on a least square estimate of a linear model. To exemplify the relevance of sensitivity analysis in this context, we can even say that sensitivity indices are to calibration, what standard statistical tests are to estimation. A detailed review on calibration and sensitivity analysis can be found in Ratto (2003), from which most of the analytical examples presented in this chapter are also taken.

The use of sensitivity analysis described here exactly fits the Setting Factors Mapping (FM), introduced in Chapter 2, in which 'we look for factors mostly responsible for producing realisations of Y in a given region'. The 'region' that we are dealing with now is the region classified as acceptable according to the categorisation criteria defined to calibrate the model, or the region of high 'fit', if some cost function or likelihood measure is used to classify the model behaviour.

In the next sections we shall discuss the two Monte Carlo calibration approaches cited above, in combination with a suitable sensitivity analysis technique. Specifically, Monte Carlo filtering is coupled with the so-called regionalised sensitivity analysis (RSA), (Hornberger and Spear, 1981); Bayesian analysis is coupled with global sensitivity analysis (Ratto *et al.*, 2001).[1]

6.2 Monte Carlo filtering and regionalised sensitivity analysis

The Monte Carlo approach to uncertainty and calibration of complex models, which came to be called regionalised sensitivity analysis (RSA), was first developed some twenty years ago within the context of environmental quality studies, (Hornberger and Spear, 1981; see also Young *et al.*, 1996; Young, 1999, and the references cited therein). In RSA a multi-parameter Monte Carlo study is performed, sampling parameters from statistical distribution

---

[1] An extension of RSA in the Bayesian framework can be found in Freer *et al.* (1996).

functions. Two tasks are required for a RSA exercise (Hornberger and Spear, 1981; Osidele, 2001):

- a qualitative definition of the system behaviour (a set of constraints: thresholds, ceilings, time bounds based on available information on the system);

- a binary classification of model outputs based on the specified behaviour definition (qualifies a simulation as behavioural, B, if the model output lies within constraints, non-behavioural, $\bar{B}$, otherwise).

Define a range for k input factors X_i $[i = 1, \ldots, k]$, reflecting uncertainties in the model and make a sufficiently large number of Monte Carlo simulations. Each Monte Carlo simulation is associated with a vector of values of the input factors. Classifying simulations as either B or $\bar{B}$, a set of binary elements are defined, distinguishing two sub-sets for each X_i: $(X_i|B)$ of m elements and $(X_i|\bar{B})$ of n elements (where $n + m = N$, the total number of Monte Carlo runs performed).

The Smirnov two-sample test (two-sided version) is performed for each factor *independently*. Under the null hypothesis that the two distributions $f_m(X_i|B)$ and $f_n(X_i|\bar{B})$ are identical:

$$H_0 : f_m(X_i|B) = f_n(X_i|\bar{B})$$
$$H_1 : f_m(X_i|B) \neq f_n(X_i|\bar{B}). \tag{6.1}$$

The test statistic is defined by

$$d_{m,n}(X_i) = \sup_y \| F_m(X_i|B) - F_n(X_i|\bar{B}) \| \tag{6.2}$$

where F are marginal cumulative probability functions, f are probability density functions.

The question answered is: 'At what significance level α does the computed value of $d_{m,n}$ determine the rejection of H_0?'

- A low level implies a significant difference between $f_m(X_i|B)$ and $f_n(X_i|\bar{B})$, suggesting that X_i is a key factor in producing the defined behaviour for the model.

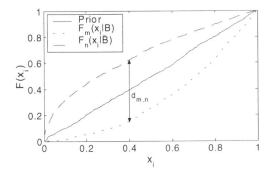

Figure 6.1 Graphical representation of the $d_{m,n}$ measure used for the Smirnov test.

- A high level supports H_0, implying an unimportant factor: any value in the predefined range is likely to fall either in B or in $\bar{B}$.

To perform the Smirnov test, we must choose the significance level α, which is the probability of rejecting H_0 when it is true (i.e. to recognise a factor as important when it is not). Derive the critical level D_α at which the computed value of $d_{m,n}$ determines the rejection of H_0 (the smaller α, the higher D_α).

If $d_{m,n} > D_\alpha$, then H_0 is rejected at significance level α. The procedure is exemplified in Figure 6.1.

The importance of the uncertainty of each parameter is inversely related to this significance level. Input factors are grouped into three sensitivity classes, based on the significance level for rejecting H_0:

1. critical $(\alpha < 1\%)$;

2. important $(\alpha \in 1\% - 10\%)$;

3. insignificant $(\alpha > 10\%)$.

6.2.1 Caveats

RSA has many 'global' properties, similar to variance based methods:

1. the whole range of values of input factors is considered;

2. all factors are varied at the same time.

Moreover, RSA classification is related to main effects of variance based methods (it analyses univariate marginal distributions).

However, no higher-order analysis is performed with the RSA approach, i.e. no attempt is made to search for interaction structure. Spear *et al.* (1994) reviewing their experience with RSA, highlighted two key drawbacks to it:

1. the success rate: the fraction of B is hardly larger than 5% over the total simulations for large models (with the number of factors $k>20$), implying a lack in statistical power;

2. correlation and interaction structures of the B subset (see also Beck's review, 1987):

    (i) the Smirnov test is a sufficient test only if H_0 is rejected (i.e. the factor is important); the acceptance of H_0 does not ensure non-importance;

    (ii) any covariance structure induced by the classification is not detected by the univariate $d_{m,n}$ statistic: for example, factors combined as products or quotients may compensate (see Example 1 below);

    (iii) bivariate correlation analysis is not revealing in many cases (see Example 2 below), i.e. the interaction structure is often far too complex for correlation analysis to be effective.

Such aspects of RSA imply that no complete assessment can be performed with RSA, since, for those factors taken as unimportant by the Smirnov test, further inspection is needed (e.g. applying global SA tools) to verify that they are not involved in higher-order interaction terms. Only after this subsequent inspection, can the relevance/unimportance of an input factor be completely assessed.

In order to address these limitations of RSA and to better understand the impact of uncertainty and interaction in the high-dimensional parameter spaces of models, Spear *et al.* (1994)

developed the computer intensive tree-structured density estimation technique (TSDE), 'which allows the characterization of complex interaction in that portion of the parameter space which gives rise to successful simulation'. Interesting applications of TSDE in environmental sciences can be found in Spear (1997), Grieb *et al.* (1999) and Osidele and Beck (2001).

In view of variance based analysis, the filtering approach has a further limitation, in that it takes into account only the output variation in the acceptable–unacceptable direction, while it ignores the variations of the output within the class of the acceptable values. In other words, an influential parameter could escape such an analysis only because it drives variation within the behavioural range.

Example 1

Let us consider the following elementary model

$$Y = X_1 X_2, \quad X_1, X_2 \sim U[-0.5, 0.5]. \tag{6.3}$$

The criterion for acceptable (behavioural) runs is $Y > 0$. Applying this criterion to a sample of X_1, X_2 gives the filtered sample shown in Figure 6.2.

When trying a Smirnov test for the acceptable/unacceptable subsets (Figure 6.3), no significance is detected for the two model parameters in driving the acceptable runs.

In this case, a correlation analysis would be helpful in highlighting the parameter interaction structure driving the model behaviour. In fact, computing the correlation coefficient of (X_1, X_2) for the B set, we get a quite high value: $\rho \cong 0.75$.

This suggests performing a Principal Component Analysis (PCA) (see, for example, Draper and Smith, 1981, for details on PCA) on the B set, obtaining the two components (the eigenvectors of the correlation matrix):

$PC_1 = (0.7079, 0.7063)$; accounting for the 87.5% of the variation of the B set.

$PC_2 = (-0.7063, 0.7079)$; accounting for the 12.5% of the variation of the B set.

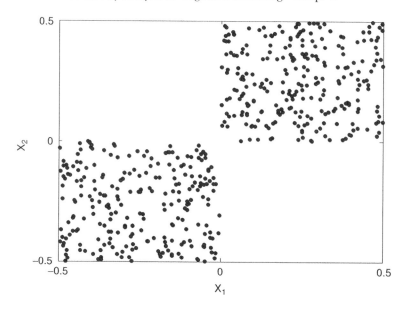

Figure 6.2   Scatter plot in the plane (X_1, X_2) of the acceptable subset B for Example 1.

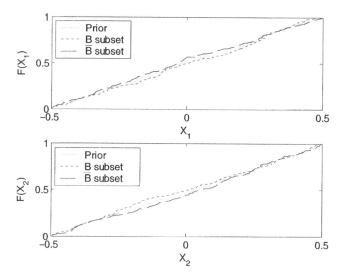

Figure 6.3   Cumulative distributions for X_1 and X_2 of the original sample (prior distribution), the B subset and the $\bar{B}$ subset. The Smirnov test in unable to highlight an effect of the model parameters in driving acceptable runs!

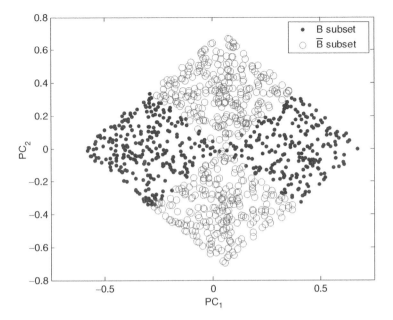

Figure 6.4   Scatter plot of the acceptable/unacceptable subsets for the principal components.

The direction of the principal component, PC_1, indicates the privileged orientation for acceptable runs.

If we make the Smirnov analysis for the principal components, we get the scatter plot in Figure 6.4 and the cumulative distributions for PC_1 and PC_2 in Figure 6.5 (original sample, B subset, $\bar{B}$ subset). Now the level of significance for rejecting the null hypothesis when it is true is very small ($< 0.1\%$), implying a very strong relevance of the linear combinations of the two input factors, defined by the principal component analysis.

Example 2

Let us consider this other model:

$$Y = X_1^2 + X_2^2, \quad X_1, X_2 \sim U[-0.5, 0.5]. \tag{6.4}$$

The criterion for acceptable runs is now: $[0.2 < Y < 0.25]$.

Plotting the B set for this case, we get the pattern in Figure 6.6.

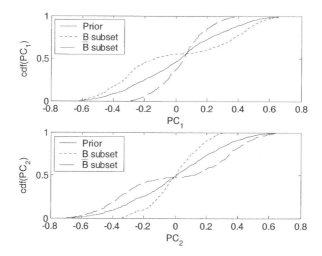

Figure 6.5   Cumulative distributions for PC_1 and PC_2 of the original sample (prior distribution), the B subset and the $\bar{B}$ subset.

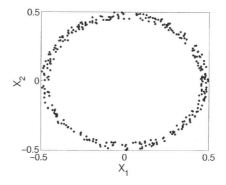

Figure 6.6   Acceptable subset B in the plane (X_1, X_2) for Example 2.

Also in this case, the Smirnov test is unable to detect any relevance for the model parameters in driving the model output behaviour (the marginal distributions are also flat in this case).

Moreover, even a correlation analysis would be ineffective for highlighting some aspects of the interaction structure driving the model acceptable runs. In fact, the empirical correlation coefficient of (X_1, X_2) of the elements of the B subset in Figure 6.6 is $\rho \approx -0.04$. This is a small value, implying that any linear transformation of (X_1, X_2) would not allow any improvement in the detection of a structure in the B subset. Moreover, although it is clear from Figure 6.6 that there is a strong coupling of (X_1, X_2) in driving the model behaviour, the correlation coefficient is absolutely ineffective in highlighting it.

6.3 Putting MC filtering and RSA to work: the problem of hedging a financial portfolio

A Monte Carlo filtering / Smirnov analysis is performed on the portfolio model shown in Chapter 3. A full description of the financial features of the portfolio under analysis can be found in Campolongo and Rossi (2002).

The goal of the exercise is to analyse the risk associated with the portfolio, identify its sources, and improve our understanding of the relationship holding between the potential loss incurred and the uncertain variables causing this loss.

The output Y of interest is the *hedging error*, which is defined as the difference between the value of the portfolio at maturity, and what would have been gained investing the initial value of the portfolio at the interest rate prevailing on the market (the market free rate). When this error is positive the bank, although failing in their hedging purpose, makes a profit. But when this error is negative, the bank faces a loss that arises from having chosen a risky investment rather than a risk-free one.

Given the problem setting, a Monte Carlo filtering / Smirnov analysis is the most appropriate here. In fact, the focus is on the identification of the factors most responsible for the model output behaviour in a region of interest, which is defined as '$Y < 0$' (negative error), where the bank is facing a loss, and in the identification of the factors most responsible for splitting the realisations of Y into acceptable–unacceptable (here 'loss–profit'). In other words, we are within the problem Setting Factors Mapping (FM), defined in Chapter 2, where the definition of an 'acceptable model behaviour' is desirable.

The hedging error, Y, depends upon a number of factors. In our analysis we decided to include only four: a factor representing the variability of the dynamics of the evolution of the interest rate through time (ε); the number of portfolio revisions to be performed

($N.\ rev.$); and the parameters a and σ of the Hull and White model of the spot rate. It is worth noting that the type of uncertainty affecting these inputs is of a different nature. The dynamics of the interest rate is an intrinsically uncertain factor, in the sense that it is totally out of the control of the analyst. In contrast, the number of portfolio revisions is a factor, which is uncertain but in a sense 'controllable' by the analyst. Any time that the portfolio is updated a cost is incurred, thus reducing the benefit derived by the update. Therefore, it is not true that the higher the number of revisions, the better the portfolio performance. There exists an optimal number of revisions that the bank may decide to carry out. Unfortunately this number is a priori unknown and therefore uncertain. Splitting the total output uncertainty into a part associated with 'uncontrollable' factors (a, σ, ε) and a part that can be reduced by optimising the input values ($N.\ rev.$) is a precious piece of information. It helps to assess the percentage of risk associated with the portfolio that is unavoidable.

The input factors' statistical distributions chosen for the uncertainty and sensitivity analysis exercise are given in Chapter 3.

A Monte Carlo filtering analysis was performed by setting as 'acceptable' output values those that are positive, i.e. corresponding to a profit for the bank. The MC sample size was set to 16 384.

The analysis was repeated for five possible scenarios resulting from five different assumptions for the values of the transaction costs. In the first scenario there are no costs. In the other four scenarios the costs are assumed to be a fixed proportion of the amount of contracts exchanged which can be either 2%, 5%, 10% or 20%. Note that the decision to treat different transaction cost values as different scenarios, rather than considering the transaction costs as an uncertain input factor taking on different possible values, was based on the fact that transaction costs are affected by natural spatial variability, varying for instance from one financial market to another. Their value is therefore unknown a priori but becomes known once the market of action has been chosen, i.e. when the model is then to be applied. The analysis is thus repeated for different scenarios to represent what happens in different markets.

First of all an uncertainty analysis was performed on the model. Results are shown in Figure 6.7, which shows the histograms of the

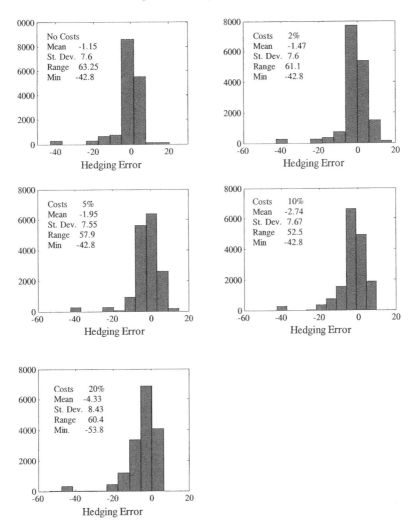

Figure 6.7   Histograms of the output distributions in each of the five transaction cost scenarios.

output distributions in each of the five transaction cost scenarios. The five histograms show that the risk of facing a loss increases as transaction costs increase. The average hedging error (loss in this case as it is negative) increases from -1.15 in the case where costs are 0 to -4.33 in the case where costs are 20%. This leads to the conclusion that the problem of hedging and offsetting financial risk gets more difficult as the costs for the transactions get higher. This

Table 6.1   Number of acceptable output values at different transaction costs.

Transaction costs	Number of acceptable values out of 16 384
0	7585
2%	6880
5%	6505
10%	5857
20%	4334

Table 6.2   Smirnov test for different values of the transaction costs.

Input factors	Transaction cost scenarios				
	0	2%	5%	10%	20%
a	0.003 3	0.003 3	0.003 3	0.003 3	0.005 3
σ	0.012 3	0.036 1	0.039 1	0.055 1	0.055 1
N. rev.	0.092 1	0.122 1	0.140 1	0.174 1	0.337 1
ε	0.432 1	0.476 1	0.504 1	0.555 1	0.534 1

result is also confirmed by the filtering analysis, which shows that the number of acceptable output values decreases as transaction costs increase (see Table 6.1).

The results of the Smirnov test are reported in Table 6.2 for several values of the transaction costs. The bold numbers indicate the level of confidence of the test results and hence the level of importance of a factor: 1 is a highly important factor, 2 is important (no factor falls in this category in Table 6.2) and 3 is not important.

As the Smirnov test is a sufficient but not necessary condition to recognise the influence of a factor, the results in Table 6.2 do not allow the conclusion that the model parameter a is irrelevant and can be fixed to its nominal base value. To complement the analysis we first computed in each scenario the correlation coefficients among factors. However this did not help, as no significant correlation values were detected.

We therefore proceeded to perform a global sensitivity analysis to the (unfiltered!) output Y, the hedging error, in order to

Table 6.3   First-order sensitivity indices.

	Tr $= 0$	Tr $= 2\%$	Tr $= 5\%$	Tr $= 10\%$	Tr $= 20\%$
a	9.1E-05	8.1E-05	8.4E-05	1.0E-04	1.3E-04
σ	9.4E-03	1.1E-02	1.3E-02	1.6E-02	1.8E-02
N. rev.	3.0E-01	3.0E-01	2.9E-01	2.6E-01	2.1E-01
ε	5.8E-02	6.8E-02	7.9E-02	9.4E-02	9.9E-02

Table 6.4   Total-order sensitivity indices.

	Tr $= 0$	Tr $= 2\%$	Tr $= 5\%$	Tr $= 10\%$	Tr $= 20\%$
a	4.4E-01	4.4E-01	4.4E-01	4.4E-01	4.8E-01
σ	4.5E-01	4.5E-01	4.5E-01	4.6E-01	5.1E-01
N. rev.	8.7E-01	8.6E-01	8.5E-01	8.3E-01	8.0E-01
ε	6.6E-01	6.6E-01	6.6E-01	6.8E-01	7.1E-01

assess the overall importance of each factor by computing its total index.

The results are shown in Table 6.3 for the first-order indices and Table 6.4 for the total-order indices.

The global sensitivity indices indicate that, although less important than the uncertainty due to the unknown interest rates dynamics, ε, or than the number of portfolio revisions performed, N. rev., the model parameter a has a non-negligible total effect ($\sim$0.4), due mostly to its interaction with other factors. Therefore its value cannot be fixed and its uncertainty should be taken into account in further studies.

The sensitivity analysis provides an encouraging insight: the uncertainty in the optimal number of revisions is the main contributor to the uncertainty in the output. As this is a 'controllable' factor, we are encouraged to carry out further analysis searching for the optimal value for this factor, thus reducing uncertainty in the analysis outcome. If this had not been the case, we would have accepted the fact that most of the uncertainty in the hedging error is due to intrinsic problem uncertainty and therefore unavoidable.

To improve our understanding of the relationship linking the number of portfolio revisions and the hedging error, we plotted histograms of the number of acceptable outputs as a function of the

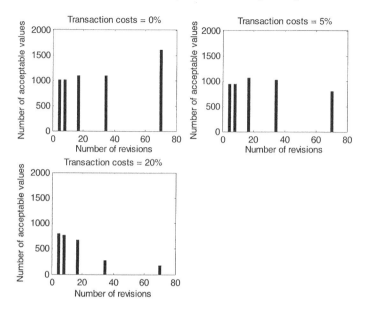

Figure 6.8 Histograms of the number of acceptable output values as a function of the number of portfolio revisions performed. Three scenarios are considered: free transaction, and costs equal respectively to 5% or 20% of the amount of contracts exchanged.

number of revisions in three scenarios (see Figure 6.8). In the first scenario, when there are no transaction costs involved, the highest percentage of acceptable values is obtained when performing the maximum number of portfolio revisions. As expected, when transaction costs are introduced, it is more appropriate to reduce the number of revisions, as evident from the extreme case when costs are up to 20% of the amount exchanged. In general, the analysis of the distribution of the acceptable values as a function of the number of portfolio revisions performed provides an indication on the optimal number of revisions that the bank should carry out in order to reduce the potential loss.

Although the test case shown here is very simple and takes into account only a limited number of uncertain input factors, it suggests that the Monte Carlo filtering/Smirnov approach is an appealing tool for financial risk management and portfolio hedging. The definition of the 'acceptable' model behaviour is particularly meaningful when addressing risk problems where the output is required to stay below a given threshold. The split of the risk into the 'not reducible' amount and its complementary part provides an indication of the actual risk faced by investors and assists them in making decisions about future hedging strategies.

The analysis also provides an indication of the type of relationship that links the input and output values, which is one of the main problems addressed by financial analysts. The MC/Smirnov analysis can be usefully complemented by a global sensitivity analysis.

6.4   Putting MC filtering and RSA to work: the Level E test case

Let us once again consider the Level E test case. In Figure 6.9 the mean and the 90% confidence bound of the dose released by the nuclear disposal obtained with Monte Carlo simulations (already shown and commented in Box 5.1 Level E).

To fix the ideas, let us concentrate on the $t = 2 \times 10^5$ yr time point. Focusing on the upper 5th percentile of the model output at

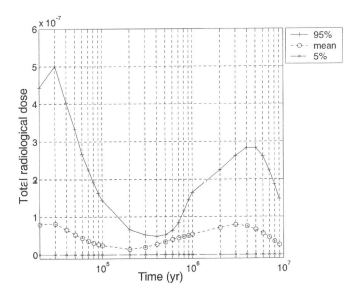

Figure 6.9   Uncertainty plot for the model output: mean and 90% uncertainy bound.

that time point and filtering the sample accordingly, the modeller can investigate which and what combination of input variables has produced such high values.

First, we can take the Smirnov test for the input factors and search for some significant separation between the two subsets obtained by the filtering procedure. Specifically, we can compare the cumulative distributions for the two subsets corresponding to the 0–95% and 95–100% percentiles of the model output at $t = 2 \times 10^5$ yr.

In Figure 6.10 the implementation of the Smirnov test is shown for all variables. For each plot, we show the significance level, α, for rejecting the null hypothesis (the higher α, the less important is the variable), together with the magnitude of the test statistic $d_{m,n}$, where $m = 1000$, $n = 19\,000$. The variables $v^{(1)}$ and W contribute significantly to producing high output values at $t = 2 \times 10^5$ yr. In particular, low values of $v^{(1)}$ and W are mainly responsible for producing high doses. $R_C^{(1)}$ and $R_C^{(2)}$ also have significant effects ($\alpha < 1\%$).

In addition to the Smirnov test on the marginal distributions, we can also analyse the correlation of the filtered sample, which sometimes is capable of revealing aspects of the interaction structure.

The 12×12 correlation matrix for the input variables calculated in correspondence with the upper 5th percentiles of the output at $t = 2 \times 10^5$ yr (see Figure 6.11) reveals an interesting correlation pattern. All significant correlation terms involve $v^{(1)}$, suggesting that this variable interacts strongly with the others, so as to produce high values of the model output. In particular, the highest correlation terms are those with the other three important variables: W (positive correlation) followed by $R_C^{(1)}$ and $R_C^{(2)}$ (negative correlation). Moreover, there are significant terms including $l^{(1)}$, $v^{(2)}$ and $l^{(2)}$. This behaviour is confirmed by the variance based analysis, in which $v^{(1)}$ has the dominant total order effect (0.91) at $t = 2 \times 10^5$ yr, while all other factors with significant total effect are the same as those highlighted by the correlation analysis.

Correlation coefficients are useful because they also suggest some qualitative way of interacting: in particular if the coefficient is positive, the pair of factors act in the model as a quotient/difference; if it is negative they act as a product/sum.

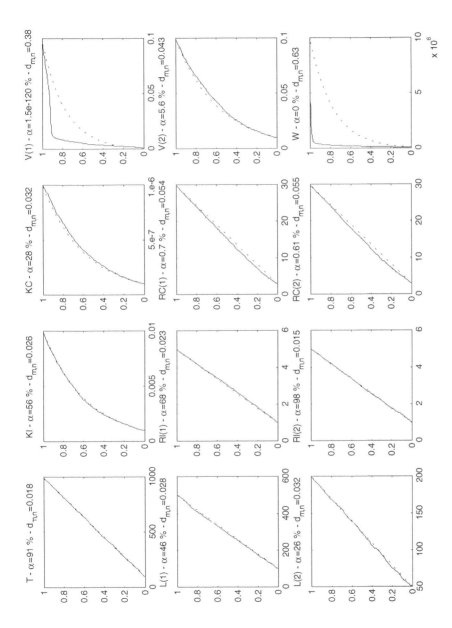

Figure 6.10 Cumulative frequency distribution for 0–95% (dashed) and 95–100% (solid) for the twelve input factors, $t = 2 \times 10^5$ yr.

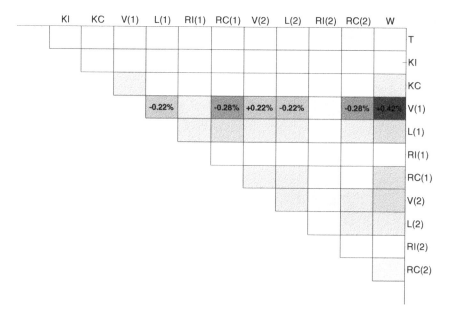

Figure 6.11   Graphical representation of the empirical correlation matrix of the input variables for the upper 5th percentile of the model output values. Darker shading indicates stronger correlation. Values are reported only for significant terms.

6.5   Bayesian uncertainty estimation and global sensitivity analysis

6.5.1   *Bayesian uncertainty estimation*

Here we go back in a simplified way to the general concepts of Bayesian analysis tools, known as Bayesian model averaging. Readers are directed, for example, to Kass and Raftery (1995) and Hoeting *et al.* (1999) for a complete and rigorous discussion on the matter.

Bayesian model averaging (BMA) is an approach to modelling in which all possible sources of uncertainty are taken into account (model structures, model parameters and data uncertainty) based on Bayesian theory.

First, let us consider a given deterministic model that has to be calibrated according to some set of observations. Assume also that the analyst has a set of prior assumptions on the model factors, expressed in terms of prior distributions $\mathrm{pr}(\mathbf{X})$.

The posterior distribution of the model factors given the data set D is, applying the Bayes chain rule,

$$\text{pr}(\mathbf{X}|D) \propto \text{pr}(D|\mathbf{X})\,\text{pr}(\mathbf{X})$$

i.e. the product of the prior by the likelihood. The likelihood function, by definition, tells us how much the experimental data support the various possible statistical hypotheses on model parameters. Analytically it is a scalar function of model parameters. For example, in a standard linear regression model where the variance of the error terms is known, the log-likelihood turns out to be proportional to the sum of the squared errors between model predictions and observations. We do not provide further details on the matter, leaving the reader to the specific literature for a comprehensive discussion. For the purposes of the present book, it is sufficient to recognise that the likelihood function is where the model structure and the data enter the problem, possibly in terms of the sums of squared errors, as in the case of a simple linear regression model.

If Y is the model output of interest and, keeping in mind that we are dealing with a deterministic model, as usual expressed as $Y = f(X_1, \ldots, X_k)$, its posterior mean and variance are expressed as follows:

$$E(Y|D) = \hat{Y} = \int f(\mathbf{X})\,\text{pr}(\mathbf{X}|D)d\mathbf{X}$$

$$V(Y|D) = \int f^2(\mathbf{X})\,\text{pr}(\mathbf{X}|D)d\mathbf{X} - \hat{Y}^2. \tag{6.5}$$

Similarly, Bayesian parameter estimates can be given in terms of posterior mean and covariance as:

$$E(\mathbf{X}|D) = \hat{\mathbf{X}} = \int \mathbf{X}\,\text{pr}(\mathbf{X}|D)d\mathbf{X}$$

$$V = COV(\mathbf{X}|D) = \int (\mathbf{X} - \hat{\mathbf{X}})(\mathbf{X} - \hat{\mathbf{X}})^{\text{T}}\,\text{pr}(\mathbf{X}|D)d\mathbf{X}. \tag{6.6}$$

As typical of Bayesian methods, BMA is an intuitively attractive solution to the problem of accounting for the different sources of uncertainty, but presents several difficulties. Among others, the integrals implicit in (6.5)–(6.6) can in general be hard to compute. Monte Carlo methods (acceptance sampling, importance sampling, see e.g. Hammersly and Handscomb, 1964; Geweke, 1999 or, better, Markov chain Monte Carlo methods (Gibbs sampler, Metropolis-Hastings, see e.g., Ripley, 1987; Geweke, 1999) have (at least partly) overcome most of the problems of efficiency and convergence of the estimation of the posterior distributions, allowing the use of Bayesian methods to grow rapidly since 1990.

If one follows the simplest Monte Carlo solution and samples the model parameters from the prior distribution $\mathrm{pr}(\mathbf{X})$, the Monte Carlo computation of integrals (6.5)–(6.6) is trivial, remembering that $\mathrm{pr}(\mathbf{X}|D) \propto \mathrm{pr}(D|\mathbf{X})\,\mathrm{pr}(\mathbf{X})$.

With $\mathbf{x}^{(i)}$ being the ith element of the N-dimensional sample of the model parameters, taken from the prior distribution, for the model output we would have:

$$E(Y|D) = \hat{Y} = \frac{\displaystyle\sum_{i=1}^{N} f(\mathbf{x}^{(i)})\,\mathrm{pr}(D|\mathbf{x}^{(i)})}{\displaystyle\sum_{i=1}^{N} \mathrm{pr}(D|\mathbf{x}^{(i)})}$$

$$V[Y|D] = \frac{\displaystyle\sum_{i=1}^{N} f^2(\mathbf{x}^{(i)})\,\mathrm{pr}(D|\mathbf{x}^{(i)})}{\displaystyle\sum_{i=1}^{N} \mathrm{pr}(D|\mathbf{x}^{(i)})} - \hat{Y}^2$$

(6.7)

while for the posterior mean and covariance of the model parameters we get:

$$E(\mathbf{X}|D) = \hat{\mathbf{X}} = \frac{\displaystyle\sum_{i=1}^{N} \mathbf{x}^{(i)}\mathrm{pr}(D|\mathbf{x}^{(i)})}{\displaystyle\sum_{i=1}^{N} \mathrm{pr}(D|\mathbf{x}^{(i)})}$$

$$\mathrm{cov}(X_l, X_j|D) = \frac{\displaystyle\sum_{i=1}^{N} \left(x_l^{(i)} - \hat{x}_l\right)\left(x_j^{(i)} - \hat{x}_j\right)\mathrm{pr}(D|\mathbf{x}^{(i)})}{\displaystyle\sum_{i=1}^{N} \mathrm{pr}(D|\mathbf{x}^{(i)})}.$$

(6.8)

Unfortunately, if the posterior distribution is concentrated relative to the prior (and this is almost always the norm), the 'rate of success' of this MC strategy is very small, i.e. most of the $\mathbf{x}^{(i)}$ have very small likelihood values, implying that all inference is dominated by a few points with large likelihood values and that the algorithm converges *very* slowly. This is the reason, why, for example, in importance sampling, one defines an 'importance sampling distribution' $j(\mathbf{x})$ allowing (hopefully!) a higher rate of success.

6.5.2 The GLUE case

The Generalised Likelihood Uncertainty Estimation (GLUE) is a simplified Bayesian approach, which is quite broadly applied in hydrological and environmental sciences, and is due to Keith Beven and co-workers (Beven and Binley, 1992). In GLUE, the Bayesian inference assumes the simplest setting described in Equations (6.7)–(6.8), i.e. in a situation in which we sample directly from the prior distributions, regardless of the efficiency problems that this might imply.

As in the more general Bayesian setting, different sets of initial, boundary conditions, model specifications or hypotheses can be considered. Based on comparing model simulations and observed responses, for example on a sum of squared scores, each set of factor values is assigned a so-called 'likelihood' of being a simulator of the system. The 'likelihood' measure as defined by Beven and co-workers does not correspond to the definition of likelihood function in estimation or Bayesian theory, but it is a qualitative measure of fit or loss function that we henceforth call the 'weighting function'. Assuming that the information set, D, consists of a single time series, Y_t, such a weighting function is typically a decreasing function of the sum of squared errors, such as:

$$\mathrm{pr}(D|\mathbf{x}^{(i)}) \equiv w(Y_t|\mathbf{x}^{(i)}) \equiv w^{(i)} \propto \exp\left(-\sigma^{(i)^2}/\sigma_{ref}^2\right), \; i = 1, \ldots, N \tag{6.9}$$

or even more simply

$$w^{(i)} \propto \left(\frac{1}{\sigma^{(i)2}}\right)^\alpha, \; i = 1, \ldots, N \tag{6.10}$$

where $\sigma^{(i)^2} = 1/2 \cdot T \sum_{t=1}^{T} (f_t(\mathbf{x}^{(i)}) - Y_t)^2$, $\mathbf{x}^{(i)}$ indicates the ith realisation of $\mathbf{X}$ drawn from its prior distribution, Y_t is the observed time series, $f_t(\mathbf{X})$ is the model evaluation at time t, T is the number of observations and N is the sample size of the Monte Carlo procedure.

Rescaling the weights such that their cumulative sum equals 1, $\sum_{i=1}^{N} w^{(i)} = 1$, yields the 'posterior weights' for the model parameters. From this the uncertainty estimation can be performed by computing the model output cumulative distribution, together with prediction quantiles.

In particular, by replacing $\mathrm{pr}(D|\mathbf{X})$ by the weighting function $w(Y_t|\mathbf{X})$ in Equations (6.7)–(6.8), we can obtain the mean and variance of the predicted variable of interest (which can be in the simplest case the prediction of $\tilde{Y}_{T+j}$ outside the observation sample) as well as the Bayesian estimates of the model parameters.

Moreover, weights can be used to estimate the posterior distributions and prediction quantiles. For example, the cumulative posterior distribution of Y_t is obtained as follows.

1. Sort the values of Y_t and store them into a new vector Y_t^*.

2. Order the w according to the sorted column vector Y_t^* and store it into a new vector w^*.

3. For all $s \in [1, N]$, define the vector of partial cumulative sums $W^{*(s)} = \sum_{j=1}^{s} w^{*(j)}$.

4. The empirical cumulative distribution function of Y_t^* is then expressed by

$$P(Y < Y^{*(1)}) = 0$$
$$P(Y^{*(s)} < Y < Y^{*(s+1)}) = W^{*(s)}, \quad \text{for } s \in [1, N-1]$$
$$P(Y > Y^{*(n)}) = 1$$

The weights are also useful for bootstrapping, i.e. re-sampling with replacement of model runs. If model runs (or input parameters) are re-sampled with a probability proportional to the weights (Russian roulette), a bootstrap sample of the posterior distribution can be obtained.

Caveats

Two main problematic aspects have to be always kept in mind when applying GLUE.

1. The definition of the weighting function is a fundamental aspect for GLUE and the uncertainty prediction can strongly depend on that definition. In a Bayesian framework, this is connected to how errors in the observations and in the model structure are represented by a statistical model. In GLUE the 'qualitative' definition of the weights, based essentially on an inverse relationship to the mean square error, makes this procedure easier and more flexible, but a bit ambiguous.

2. The sampling strategy of GLUE has very poor efficiency properties, as discussed in the previous section, which can make the statistical properties of the GLUE inference poorly significant. The use of importance sampling could be a first step in the direction of improving efficiency, without introducing too much complication in the methodology (see also Young and Romanowicz, 2003, for a further discussion on this matter and Romanowicz *et al.*, 1994, for a further discussion on GLUE and Bayesian analysis).

The GLUE methodology has been applied to a variety of environmental prediction problems, such as rainfall-runoff modelling (Beven and Binley, 1992; Beven, 2001), flood inundation prediction (Romanowicz and Beven, 1998; Beven *et al.*, 2000) and air pollution modelling (Romanowicz *et al.*, 2000).

6.5.3 *Using global sensitivity analysis in the Bayesian uncertainty estimation*

As discussed at the beginning of this chapter, the aim of applying sensitivity analysis in the Bayesian framework is to address the problem of describing the acceptable parameter structure, i.e. to specify the parameter calibration (Ratto *et al.*, 2001). There are at least two possibilities for performing a sensitivity analysis: on the model output itself or on the likelihood. The calibration issue is addressed by performing a global sensitivity analysis of the likelihood (or of the weighting function for GLUE).

Taking the global SA of the likelihood, means decomposing the *variance* of the likelihood over the model parameter space, i.e. we are asking ourselves which parameters drive most of the variation of the likelihood. Is this a significant analysis? In other words, *does the analysis of the variance give information on which model parameters mainly drive the goodness of the model?* In order to answer this question let us consider the following example.

As described in Chapter 5, sensitivity is computed through variances of conditional expectations. If we are analysing the likelihood function, the main effect is given by

$$S_i = V_i / V = V[E(\text{pr}(D|X_1, \ldots, X_k)|X_i)] / V \qquad (6.11)$$

This quantity measures the variation of the univariate function $E[\text{pr}(D|X_1, \ldots, X_k)|X_i] \equiv \text{pr}(D|X_i)$ around the unconditional mean $\text{pr}(D)$. The latter expression is given by $\text{pr}(D) = \int \text{pr}(D|\mathbf{X}) \, \text{pr}(\mathbf{X}) d\mathbf{X}$ and is called the 'integrated likelihood' in the Bayesian literature. Such a quantity is a *measure of the goodness of a model*. Accordingly, we can define $\text{pr}(D|X_i)$ as the conditional integrated likelihood.

If the main effect is high, the *conditional* integrated likelihood $\text{pr}(D|X_i)$ will have a strong pattern, i.e. there will be values of X_i for which $\text{pr}(D|X_i)$ is *significantly smaller* than $\text{pr}(D)$ and other values for which $\text{pr}(D|X_i)$ is *significantly larger* than $\text{pr}(D)$. This means that if we were allowed to tune (or fix) the values of X_i, we would be able to let the integrated likelihood increase (i.e. increase the goodness of the model). The same holds if groups of factors are considered, allowing one to identify interaction structures for the acceptable behaviour.

Developing this reasoning further, it can be demonstrated that from the computation of main effects and total effects, we can classify the model factors in terms of necessary and sufficient conditions as follows:

1. factors with a high main effect: such factors affect model goodness singularly, independently of interaction (necessary and sufficient condition);

2. factors with a small main effect but high total effect: such factors influence the model goodness mainly through interaction (necessary and sufficient condition);

3. factors with a small main and total effect: such factors have a negligible effect on the model goodness and can be ignored in the further inspection of the behavioural regions (necessary and sufficient condition).

The first class of factors can also be detected with the other methodological approaches (RSA, scatter plots, regression analysis), while the second class can in general be detected only using global SA or other advanced techniques such as the tree-structured density estimation of Spear *et al.* (1994). Moreover, as a general rule, a large difference between main and total effects implies that the model is over-parameterised.

Finally, if group effects are analysed:

4. for groups of factors having a high group effect: such a groups affect model goodness singularly, independently of interaction with parameters outside that group (necessary and sufficient condition).

In conclusion, global SA is able to highlight much more complex interaction structures than classical analysis such as PCA or co-variance analysis. The main limit is that global SA can identify the key parameters for the interaction structure, but it gives no 'topological' information on the relevant behavioural zones. However, global SA is able to give *quantitative* and synthetic information, in terms of *necessary and sufficient* conditions, which facilitates the subsequent inspection of the behavioural regions. Since a global SA sample is just like any other Monte Carlo sample used for Bayesian uncertainty estimation, we get additional useful information at virtually no additional cost in terms of number of model evaluations.

Any other tool adopted to represent the interaction structure – from correlation coefficients to Principal Component Analysis to Bayesian networks to tree-structured density estimation – will conform to the general features identified by global SA.

6.5.4 Implementation of the method

In practice, the way of combining global SA and Bayesian analysis is straightforward. It is necessary that the sample generated for the Bayesian analysis is also designed for the computation of variance-based sensitivity indices. In this way, by applying the same set of model runs, predictive uncertainty can be estimated and sensitivity indices computed. This is particularly simple in the GLUE case, since the samples are drawn directly from the prior distributions and usually no correlation terms in the prior distributions are considered in GLUE, so that the sampling designs for orthogonal inputs such as Sobol' or FAST can be directly applied.[2]

6.6 Putting Bayesian analysis and global SA to work: two spheres

Let us consider the 'Two spheres' example, introduced in Chapters 2 and 3. Let us perform a Monte Carlo analysis, sampling the six input factors from normal prior distributions $N(0, 0.35)$. We want to analyse the following function, as if it were a black-box function

$$f(X_1, \ldots, X_6) = -\left(\sqrt{X_1^2 + X_2^2 + X_3^2} - R_1\right)^2 / A_1$$
$$-\left(\sqrt{X_4^2 + X_5^2 + X_6^2} - R_2\right)^2 / A_2 \quad (6.12)$$

which, as explained in Chapters 2 and 3, represents a likelihood or a weighting function obtained in an estimation/calibration procedure, e.g. it can be seen as the kernel of a weighting function obtained using a GLUE approach: $f(X_1, \ldots, X_6) \propto w(D| X_1, \ldots, X_6)$.[3]

---

[2] If importance sampling techniques were used (in order to improve the efficiency of GLUE), the sample would surely be non-orthogonal, and sampling designs of variance based global SA methods for non-orthogonal inputs could still be used straightforwardly; while in the case of MCMC techniques, the use of classical sampling designs of variance based global SA techniques would be problematic. In such cases, approximated estimation tools are available, as described in Ratto and Tarantola (2003).

[3] The scale of f is actually a log-likelihood scale (see negative values in Figures 6.12–6.13).

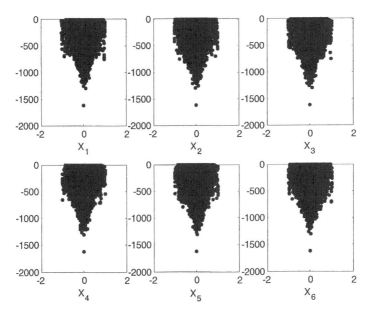

Figure 6.12   Scatter plots of $f(X_1, \ldots, X_6)$.

Scatter plots for the plain Monte Carlo runs are shown in Figure 6.12. We can see that there is a strong pattern (i.e. main effect) of factors in driving 'bad' runs (small values of f for central values of parameters), but the maximum seems flat.

A first step for inspecting the acceptable parameter regions could be to proceed with a filtering. We filtered runs according to the rule $f > (-200)$ and plotted the results in Figure 6.13. In the scatter plots, no pattern is visible and nothing can be said regarding an optimal subset from the marginal distributions of the six-dimensional parameter space. In other words, nothing can be said in terms of first-order effects.

Another 'classical' approach would be to look at some correlation structure. This can be done by analysing the correlation matrix of the filtered sample: no significant correlation term is detected. As shown in Table 6.5, the largest correlation term is, in absolute value, 0.0254.

Consequently, a PCA would also give no useful result in describing the behavioural structure of the factors, being based on

Table 6.5   Correlation coefficients of the sample filtered according to the rule $f > (-200)$.

	X_1	X_2	X_3	X_4	X_5	X_6
X_1	1.0000	−0.0031	−0.0112	−0.0180	−0.0195	−0.0129
X_2	−0.0031	1.0000	−0.0133	−0.0160	−0.0086	−0.0254
X_3	−0.0112	−0.0133	1.0000	−0.0191	−0.0010	−0.0239
X_4	−0.0180	−0.0160	−0.0191	1.0000	−0.0056	−0.0220
X_5	−0.0195	−0.0086	−0.0010	−0.0056	1.0000	−0.0127
X_6	−0.0129	−0.0254	−0.0239	−0.0220	−0.0127	1.0000

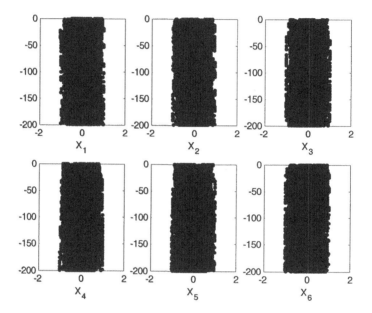

Figure 6.13   Scatter plots of $f(X_1, \ldots, X_6)$ filtering runs according to the rule $f > (-200)$.

the correlation matrix (see e.g. Ratto *et al.*, 2001, for the use of PCA in identifying structures in the behavioural sets).

A further possibility would be to consider the two-dimensional projections of the filtered sample (Figure 6.15) and compare it to the original sample (Figure 6.14); also in this case, no structure is detected.

A further resource for analysing the behavioural structure is given by global SA. Sensitivity indices of first, second and third order are shown in Figure 6.16 and Figure 6.17.

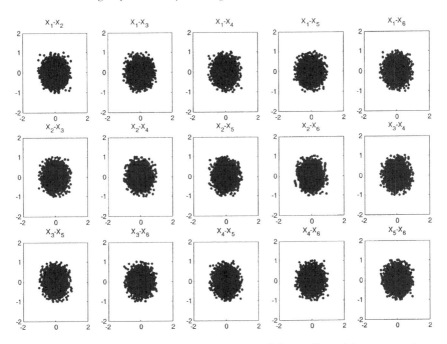

Figure 6.14   Two-dimensional projection of the unfiltered input sample.

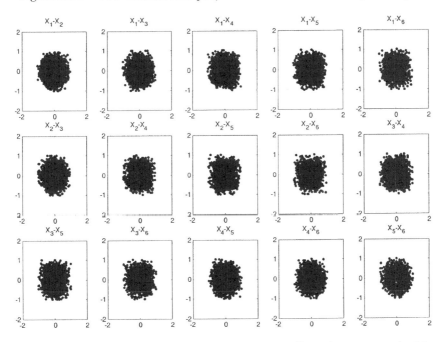

Figure 6.15   Two-dimensional projections of the filtered input sample. No structure appears, with respect to the unfiltered input sample.

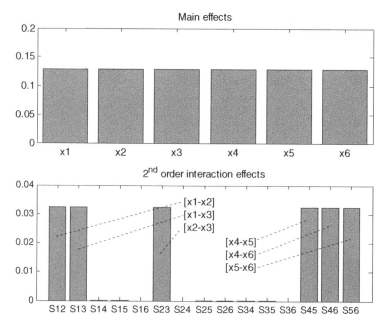

Figure 6.16   Main effects (upper panel) and second-order effects of the two-spheres problem (Sobol' estimates).

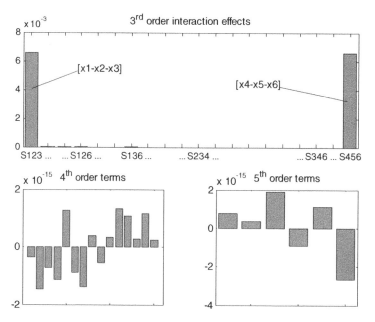

Figure 6.17   Third-, fourth-, and fifth-order effects (Sobol' estimates).

Main effects (Figure 6.16, upper panel). There are significant main effects for all factors, mainly due to the effect of single factors in driving 'bad' runs (connected to small values of f in Figure 6.12).

Second-order effects (Figure 6.16, lower panel). Relevant second-order effects can be noticed for the sub-sets involving the groups [X_1, X_2, X_3] and [X_4, X_5, X_6]. This is another major difference with standard correlation or projection analysis. From the definition of the objective function, it is clear that first-, second- and third-order interaction terms should be present. It is very important that global SA is able to detect second-order terms, while PCA or correlation analysis are not, implying that global SA is more powerful in identifying interaction structures.

Third-order effects (Figure 6.17, upper panel). Only two peaks of third-order effects are detected, corresponding to the groups [X_1, X_2, X_3] and [X_4, X_5, X_6]. Again, this is a very clear example of the capabilities of the variance decomposition in highlighting interaction structures.

Higher-order effects (Figure 6.17, lower panel). Estimates of the higher-order effects are almost null (10^{-15}), so it is evident that no interaction larger than three-dimensions is present (theoretical values are exactly zero!).

Finally, the global SA analysis can be concluded considering the estimates of the third-order closed effects of the groups [X_1, X_2, X_3] and [X_4, X_5, X_6]:

$$S_{123}^c = S_1 + S_2 + S_3 + S_{12} + S_{13} + S_{23} + S_{123} = 0.5$$
$$S_{456}^c = S_4 + S_5 + S_6 + S_{45} + S_{46} + S_{56} + S_{456} = 0.5$$

The estimates of such third-order closed effects sum exactly to 1, clearly implying that interaction occurs within each subset but not across them.

From global SA it is possible to conclude that the interaction structure yielding behavioural parameter subsets is given by two separated subsets in three-dimensions: [X_1, X_2, X_3] and [X_4, X_5, X_6].

Global SA is unable to show exactly the spherical symmetry of the problem, but does give a hint of it, highlighting the two main subsets of factors. No simple method would in any case be able to provide such detailed information on spherical symmetry.

Readers can imagine what kind of difficulties can be encountered by a modeller in trying to describe the acceptable model behaviour in a calibration exercise of *real* black-box system, i.e. characterised by a complex computational model, which is compared with a set of observations.

6.7 Putting Bayesian analysis and global SA to work: a chemical experiment

Let us consider the simple chemical system introduced in Chapter 3, consisting of the observation of the time evolution of an isothermal first-order irreversible reaction in a batch system $A \rightarrow B$. The analyst wants to calibrate a simple kinetic model considering the pseudo-experiment shown in Figure 6.18.

There were three factors to be considered for the calibration study $\mathbf{X} = [k_\infty, E, y_B^0]$. A sample of size 2048 was generated to estimate sensitivity indices (first and total effect). Two model outputs have been considered: the physical output $y_B(t)$ and the weighting function, based on the mean squared error σ^2:

$$ f\left(x_1^{(i)}, x_2^{(i)}, x_3^{(i)}\right) = w^{(i)} \propto \left(\frac{1}{\sigma^{(i)2}}\right)^\alpha , \quad i = 1, \ldots, N \quad (6.13) $$

with $\alpha = 1, 4$. This weighting function is obtained from a GLUE type analysis (Bayesian simplified approach). By increasing α, we

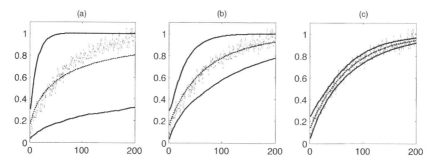

Figure 6.18 Experimental time series and 5% and 95% confidence bounds for the output y_B: (a) no weighting function used; (b) using the weighting function $1/\sigma^2$; (c) using the weighting function $(1/\sigma^2)^4$.

give a much higher weight to good runs, while most runs are classified as 'unlikely'.

6.7.1 Bayesian uncertainty analysis (GLUE case)

In Figure 6.18 the confidence bound (5% and 95%) of the output $y_B(t)$ is shown, for three cases:

(a) no weighting function is applied and the uncertainty bound is given purely by the propagation of the prior distributions of the model parameters;
(b) applying the weighting function (6.13) with $\alpha = 1$;
(c) applying the weighting function (6.13) with $\alpha = 4$.

This is an example of the use of GLUE for the prediction uncertainty. The effect of using the weights for constructing the uncertainty bounds is clear from the three plots. Mean values and confidence bounds change drastically when different types of weighting functions are applied. This allows one to perform an uncertainty analysis using data, without any true estimation step. On the other hand, this also makes clear the arbitrariness of the definition of the weighting function, compared with rigorous Bayesian analysis.

6.7.2 Global sensitivity analysis

In Figure 6.19 scatter plots are shown for the weighting function with $\alpha = 1$ vs. the three factors. Sensitivity indices are shown in Figure 6.20 for the physical output $y_B(t)$ and in Figure 6.21 for the weighting function. Scatter plots provide the same type of information as main effect sensitivity indices. In fact, the conditional variance defining main effects can be 'visualised' in scatter plots: a high main effect corresponds to a clear pattern in scatter plots.

Analysis of the physical output
Sensitivity indices have a trend in time where the initial condition y_B^0 is important for the very initial time period, while the factors of the chemical rate constant prevail for the rest of the simulation. The sum of the first-order indices is never less than 0.86. By

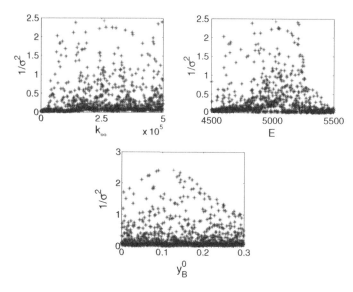

Figure 6.19   Scatter plots of the weighting function $1/\sigma^2$ vs. k_∞, E and $y_B{}^0$.

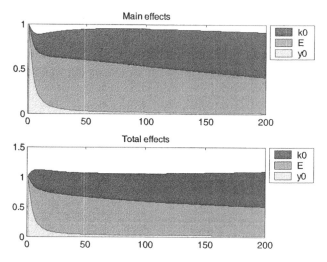

Figure 6.20   Sobol' sensitivity indices for the output $y_B(t)$ [cumulative plot].

considering the total effect sensitivity indices, a very slight increase in the absolute values with respect to the first-order sensitivity indices is detected. This implies that little interaction is revealed by the analysis of the physical output, which simply singles out the importance of *both* kinetic factors.

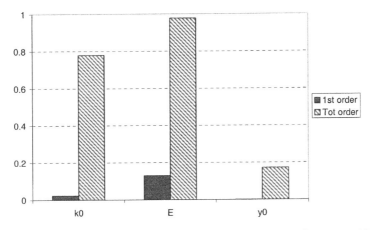

Figure 6.21 Sobol' sensitivity indices for the weighting function $(1/\sigma^2)$.

Analysis of the weighting function

In this case, the dependence over time is eliminated, making the analysis more synthetic and strongly changing the sensitivity behaviour. By considering the scatter plots, no clear trend can be seen for any of the three factors. Also, applying the Smirnov test (not shown here), no appreciable separation in the factor distribution is detected, when filtering runs with high/small weights. This is reflected in the first-order sensitivity indices, which are much smaller than the main effects for the physical output. From the analysis of the main effect (Smirnov, scatter plots, first-order indices) we can conclude that no single factor drives the model to be more 'behavioural' and that interaction mainly characterises model calibration. On the basis of the main effect, it is not possible to get any information about the interaction structure.

By analysing the total effect indices, very high sensitivity is detected for the chemical kinetics factors, implying that the behavioural runs are driven by an interaction between them. On the other hand, the influence of the initial condition is also small in terms of the total effect.

Conclusions drawn from global SA

From these results, one may conclude the following.

1. The initial condition can be judged as unimportant, as it has the smallest total effect (sufficient and necessary condition).

2. The chemical rate factors mainly drive the model fit to the experimental data, as they have the highest main and total effects (sufficient and necessary condition).

3. On the other hand, the chemical rate factors cannot be precisely estimated (unidentifiability), as the absolute values of the first-order indices are small, leaving the main contribution to the output variance to interaction terms.

4. The high difference between main and total effects implies that the model is over-parameterised.

Of particular interest is the relationship singled out in items (3) and (4) between (i) the difference between total- and first-order sensitivity indices, (ii) the indeterminacy of the optimisation (estimation) problem, and (iii) the interaction structure of the input factors in the posterior distribution after conditioning to the observations.

SA on model output and on its weighting function: what differences?
By comparing results in the previous sections, it is evident that the input–output structure is much more complicated when using the weights than when considering the physical output. In particular, we should generally expect that the weighting function is non-monotonic with respect to the input factors and that more interactions are reflected by the use of the weights. This implies some restriction as far as the SA tools to be applied: specifically, only variance-based methods are suitable, since they are model free, they are able to deal with non-monotonic behaviour and to reveal interaction terms.

6.7.3 Correlation analysis

In this simple case, a correlation analysis of the posterior joint distribution of the model parameters would be useful. Correlation coefficients allow one to evaluate the pair-wise interaction structure.

The matrix shown in Table 6.6 gives the posterior correlation structure. The posterior distribution confirms the interaction

Table 6.6 Estimate of the correlation matrix of the posterior joint distribution.

	k_∞	E	$y_B{}^0$
k_∞	1	0.6682	−0.0387
E	0.6682	1	0.0901
$y_B{}^0$	−0.0387	0.0901	1

between the kinetic factors highlighted by the global SA. When high values of the correlation coefficients are detected, they also suggest a way to reduce the input factor space. In particular, if the coefficient is positive, the couple of factors acts in the model as a quotient/difference, if it is negative they act as a product/sum. In the case under analysis, the positive sign correctly reveals the quotient interaction of k_∞ and E. This is a clarification of what we claimed in the previous paragraphs: global SA allows a general, quantitative, model free identification of basic features of the interaction structure. On the other hand, it does not allow a complete analytical representation of such a structure. Such a representation can be drawn by applying other tools, which, in turn, require the introduction of more stringent assumptions about the interaction structure and have a less general applicability. In all cases, such representations confirm global SA results (in this case the interaction between the kinetic factors) and global SA, therefore, is a 'common denominator' to them.

6.7.4 Further analysis by varying temperature in the data set: fewer interactions in the model

Let us now consider the same chemical system, but assume that nine sets of observations are available at nine different temperatures: in particular, we considered five measurements at each temperature for a total of 45 observations. The new pseudo-experimental data are shown in Figure 6.22. It is assumed that the temperature of each observation is known so that the model always contains three factors for calibration. The weighting function is always the inverse of the mean square difference between the model and experiments over the nine time series (equation 6.13 with $\alpha = 1$).

Table 6.7   Sensitivity indices obtained with the new observation data set (Figure 6.22).

	First order	Tot. order
k_∞	0.63072	0.68821
E	0.07971	0.11696
y_B^0	0.29506	0.3192

Table 6.8   Correlation matrix of the posterior joint pdf obtained by performing the Bayesian and global SA analysis with the new observation data set (Figure 6.22).

	k_∞	E	y_B^0
k_∞	1	0.0366	−0.0432
E	0.0366	1	0.0079
y_B^0	−0.0432	0.0079	1

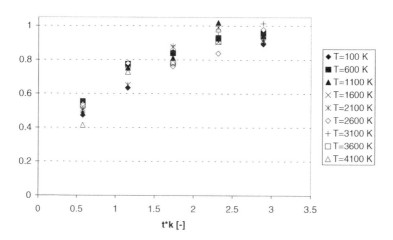

Figure 6.22   Experimental data at nine different temperature versus dimensionless time-scale.

Sensitivity indices for the weighting function are shown in Table 6.7. The correlation matrix under the posterior joint pdf is shown in Table 6.8.

As expected, when the temperature range of the different experimental measurements is varied significantly, the interaction between the kinetic factors is strongly reduced. Correspondingly the absolute values of the first-order sensitivity indices become much larger, summing almost to one. Under the particular operating conditions chosen, the influence of the two kinetic factors does not split uniformly, but concentrates on k_∞. This is not a general result. What has to be expected in general is the decrease in the interaction of the model as a whole. This means that the 'posterior' pdf (probability distribution function) structure can be described in elementary terms as a summation of first-order effects.

Also the correlation structure is now very weak, confirming that by changing the data set, the effect of the kinetic factors is de-coupled. This also exemplifies that a model can be over-parameterised or not, according to the evidence with which it is compared, implying that the calibration and sensitivity analysis exercise is also useful when the same mathematical model is used to describe different realities.

Finally, parameter identifiability can also be more precisely assessed, as in the classical estimation problems. In the present case, the Arrhenius pre-exponential factor will be very well estimated, while the activation energy is not well determined, not because of under-determination, but because it does not significantly affect the 'objective' function.

6.8   Caveats

The performance of a global SA to the weighting/likelihood functions gives a full interaction structure, which obviously is not the same as a functional representation.[4] The latter is something additional with respect to the performance of a global SA and can, in some cases, be a formidable task. This task usually requires the use of computationally intensive methods and/or the formulation of hypotheses about the interaction structure and the introduction of a certain degree of arbitrariness for such a representation.

---

[4] An example of functional representation is the High Dimensional Model Representation, see Box 2.5 in Chapter 2.

On the other hand, the application of a global SA provides a quantitative evaluation about fundamental aspects of the calibration problem, such as:

- which factors are important for calibration, i.e. which are somehow conditioned by observations;

- the degree of complexity of the interaction structure;

- which factors are involved in the interaction structure.

Such information has a general validity, since it is obtained without assumptions about the model structure and/or the error structure. So, global SA reveals the fundamental properties of such a structure, which are common to any more detailed representation and which are not affected by any 'modeller's prejudice'.

7 HOW TO USE SIMLAB

7.1 Introduction

SIMLAB is didactical software designed for global uncertainty and sensitivity analysis. These analyses are based on performing multiple model evaluations with probabilistically selected input factors, and then using the results of these evaluations to determine (1) the uncertainty in model predictions and (2) the input factors that gave rise to this uncertainty. To use SIMLAB the user performs the following operations.

1. Select a range and distribution for each input factor. These selections will be used in the next step for the generation of a sample from the input factors. If the analysis is primarily of an exploratory nature, then quite rough distribution assumptions may be adequate.

2. Generate a sample of elements from the distribution of the inputs previously specified. The result of this step is a sequence of sample elements.

3. Feed the model with the sample elements and produce a set of model outputs. In essence, these model evaluations create a mapping from the space of the inputs to the space of the results. This mapping is the basis for subsequent uncertainty and sensitivity analysis.

4. Use the results of model evaluations as the basis for uncertainty analysis. One way to characterise the uncertainty is with a mean value and a variance. Other model output statistics are provided.

Sensitivity Analysis in Practice: A Guide to Assessing Scientific Models A. Saltelli, S. Tarantola, F. Campolongo and M. Ratto © 2004 John Wiley & Sons, Ltd. ISBN 0-470-87093-1

5. Use the results of model evaluations as the basis for sensitivity analysis.

This chapter gives an overview of the functionalities of the software. For a deeper understanding and practical use, the reader is referred to the on-line manual.

7.2 How to obtain and install SIMLAB

SIMLAB can be downloaded at the URL of this book http://www.jrc.cec.eu.int/uasa/primer-SA.asp or, directly, at the SIMLAB URL: http://www.jrc.cec.eu.int/uasa/prj-sa-soft.asp. To install the software the user has to be provided with a personal licence number. To obtain the licence number, please email: stefano.tarantola@jrc.it. After installing SIMLAB, the user has to set the following PC options.

1. Start the *Control Panel,* click on *Regional Options* and then on *Numbers.*

2. Set the Decimal Symbol to 'Dot' and Digit Grouping Symbol to 'blank'.

7.3 SIMLAB main panel

At the start SIMLAB displays the main panel (Figure 7.1); this panel is logically divided in three frames:

1. The *Statistical Pre Processor module*: generates a sample in the space of the input factors.

2. The *Model Execution module*: executes the model for each point in the sample of input factors.

3. The *Statistical Post Processor module*: performs the uncertainty and sensitivity analysis.

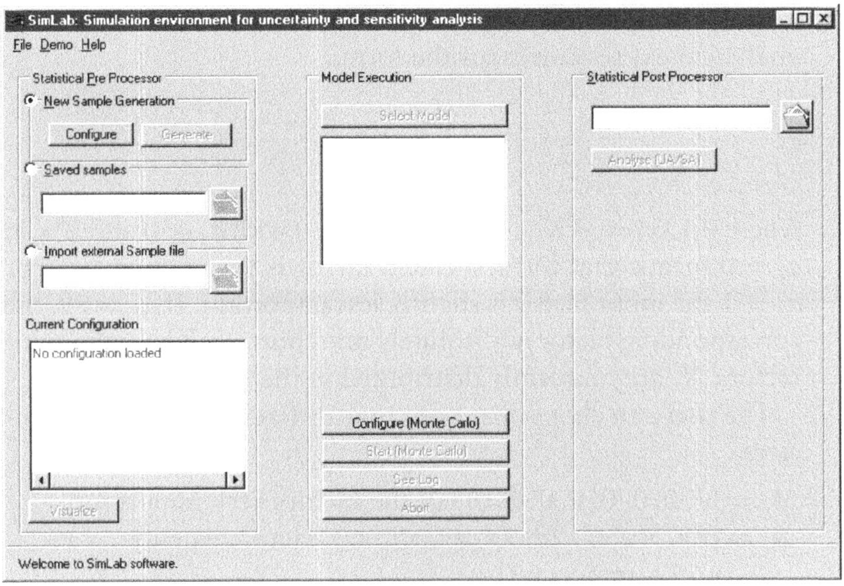

| Statistical Preprocessor | Model execution | Statistical Postprocessor |
| module | module | module |

Figure 7.1   SIMLAB main panel.

The SIMLAB main panel provides a special configuration called *Demo mode* that contains some test functions. The user can test different sampling strategies for the same built-in demo model. Demo files will not be overwritten.

The codes of the test cases shown in this book are available as executable files in *.\SIMLAB\models*. This enables the user to test SIMLAB on these functions. SIMLAB provides four different test functions in the *Demo mode*:

1. *Linear model.* Three input factors, X_1, X_2, X_3, are defined as uniform distributions varying over a specific range:

$$X_1 \sim U(0.5, \ 1.5)$$
$$X_2 \sim U(1.5, \ 4.5)$$
$$X_3 \sim U(4.5, \ 13.5).$$

The output variable Y is simply the sum of the three input factors.

2. *G-function of Sobol'*. This is a non-monotonic function whose analytical expression takes the form:

$$Y = \prod_{j=1}^{k} g_j(X_j)$$

where $g_j(X_j) = (|4X_j - 2| + a_j)/(1 + a_j)$ with $a_j \geq 0$ and $k = 8$. $a_j = 0$ means that the associated factor is very important. For $a_j = 1$ the input factor is slightly less important. If $a_j = 99$, the corresponding factor is absolutely unimportant. The eight input factors X_j are uniformly distributed in the range (0,1).

The user can choose between four different sets of a_j parameters:

$a_j = \{0, 0, 0, 0, 0, 0, 0, 0\}$ [all the factors very important]
$a_j = \{99, 99, 99, 99, 99, 99, 99, 99\}$ [factors equally non − important]
$a_j = \{0, 1, 4.5, 9, 99, 99, 99, 99\}$ [in decreasing order of importance]
$a_j = \{99, 0, 9, 0, 99, 4.5, 1, 99\}$ [in random order of importance]

3. *Ishigami function*. Another non-monotonic function with three input factors X_1, X_2, X_3. The factors are uniformly distributed in $(-\pi, \pi)$. The model is:

$$Y = \sin X_1 + A \sin^2 X_2 + B X_3^4 \sin X_1.$$

The parameters A and B have values $A = 7$ and $B = 0.1$. The main peculiarity of this model is the dependence on X_3, which has no addictive effect on Y but interacts only with X_1 (see also Box 2.5 in Chapter 2).

4. *Level E*. The Level E model has been treated in detail in this book (Chapters 3, 5 and 6). It is a computer code used in safety assessment for nuclear waste disposal. It predicts the radiological dose to humans over geological time scales due to the underground migration of radionuclides from a nuclear waste disposal site through a system of natural and engineered barriers. The core of the model is a set of partial differential equations that describe the migration of four nuclides through two

geosphere layers characterized by different hydro-geological properties (see Chapter 3). The processes being considered in the model are radioactive decay, dispersion, advection and chemical retention between the migrating nuclides and the porous medium. The model has a total of 33 factors, 12 of which are taken as independent uncertain factors for SA, see Table 3.5.

The user will exit the *Demo mode* by choosing the menu item *Exit Demo mode* in the *Demo menu* or pressing the button *Exit Demo mode* located on the SIMLAB main panel. The demo configuration will be discarded, so the user will have to save it under a different name if any changes have been made.

7.4   Sample generation

The first step in the sample generation phase is to select ranges and distributions (probability distribution functions, pdfs) for the input factors. This selection makes use of the best information available on the statistical properties of the input factors. In some instances, it is possible to get empirical estimates of pdfs from available underlying data for the input factors. The effort put in at this stage is related to the scope of the analysis: if the analysis is at an exploratory stage, then rather crude pdfs could be adequate: the cost of getting better pdfs may be relatively high.

A set of twelve types of pdfs is available in SIMLAB. A complete list of the distributions and a description of the corresponding panels is given in Appendix C of the on-line manual.

The second step is to select a sampling method from among the available ones. The sampling techniques available in SIMLAB are FAST, Extended FAST, Fixed sampling (a predetermined sequence of points), Latin Hypercube, replicated Latin Hypercube, Morris, Quasi-random LpTau, Random and Sobol' (see below). SIMLAB visualises only those techniques that are suitable for the current factor configuration. Note also that the choice of sampling method has implications on what type of sensitivity analysis the user is able to perform later (for example, if the user selects FAST sampling, he cannot perform the analysis using the method of Morris).

The third step is the actual generation of the sample from the input pdfs. The generated sample can be visualised using scatter plots, cobwebs, histograms and tables. The user can also look at the sample correlation.

7.4.1   FAST

The classical FAST method estimates the first-order effects. The extension of FAST computes first-order effects and total effects. This technique can also be used by grouping sub-sets of factors together. The FAST method can be used with a set of orthogonal factors. The algorithm is based on a transformation that converts a multidimensional integral over all the uncertain model inputs into a one-dimensional integral. Specifically, a search curve, which scans the whole parameter space, is constructed in order to avoid the multidimensional integration. A decomposition of the Fourier series representation is used to obtain the fractional contribution of the individual input variables to the variance of the model prediction.

7.4.2   Fixed sampling

The generation of the sample is completely controlled by the user, who decides where to select the sample points within the sample space.

7.4.3   Latin hypercube sampling (LHS)

This is a particular case of stratified sampling. LHS performs better than random sampling when the output is dominated by a few components of the input factors. The method ensures that each of these components is represented in a fully stratified manner, no matter which components might turn out to be important. LHS is better than random sampling for estimating the mean and the population distribution function. LHS is asymptotically better than random sampling in that it provides an estimator (of the expectation of the output function) with lower variance. In particular, the closer the output function is to being additive in its input variables, the more reduction in variance. LHS yields biased estimates of the variance of the output variables.

7.4.4   *The method of Morris*

The guiding philosophy of the Morris method is to determine which factors may be considered to have effects, which are negligible, linear and additive, or non-linear or involved in interactions with other parameters. The sensitivity measures provided in SIMLAB are μ^* and σ (see Chapter 4). The experimental plan is composed of individually randomised 'one-factor-at-a-time' experiments, in which the impact of changing the value of each of the chosen factors is evaluated in turn.

The number of model executions is computed as $r(k+1)$, where r is the number of trajectories (sequences of points starting from a random base vector in which two consecutive elements differ only for one component) and k, the number of model input factors.

For each factor, the Morris method operates on selected levels. These levels correspond to the quantiles of the factor distribution. In particular:

For four levels, the 12.50th, 37.50th, 62.50th and 87.50th quantiles are taken.

For six levels, the 8.33th, 25.00th, 41.66th, 58.33th, 75.00th, 91.66th quantiles are taken.

For eight levels, the 6.25th, 18.75th, 31.25th, 43.75th, 56.25th, 68.75th, 81.25th and 93.75th quantiles are taken.

The method of Morris can only be used with a set of orthogonal factors.

7.4.5   *Quasi-Random LpTau*

The method generates uniformly distributed quasi-random sequences within the hypercube $\Omega = \{[0; 1] \times [0; 1] \times \ldots\}$ of unit volume that have the property of minimising discrepancy. The sample looks like a quasi-regular grid of points that are located so that there exists a direct proportionality between any hypercube of volume $V < 1$ and the number of sample points within that hypercube, whatever hypercube is considered in Ω. From the unit hypercube, the sample is then automatically transformed to the actual parameters values using the marginal distributions defined by the user.

The method generates sets of orthogonal input factors. The number of input factors cannot exceed 51.

7.4.6 Random

With this method a sample of the desired dimension is generated from the marginal distributions. Random sampling is also referred to as pseudo random, as the random numbers are machine-generated with deterministic process. Statistically, random sampling has advantages, as it produces unbiased estimates of the mean and the variance of the output variables.

7.4.7 Replicated Latin Hypercube (r-LHS)

The sample in r-LHS is generated by replicating r times a base sample set created using the LHS algorithm (see Chapter 5 for details). The r-LHS is used to estimate importance measures (also called correlation ratios, in other words $V[E(Y|X_i)]/V(Y)$). Given the high computational effort required, this technique is employed in SIMLAB only for non-orthogonal input. r-LHS is used in association with either the Iman–Conover rank correlation method or the Stein method (see below for more information on how to induce dependencies in the input factors).

7.4.8 The method of Sobol'

The method prepares a sample for subsequent use in the estimation of the Sobol's sensitivity indices. The users can select up to what order they want to estimate the sensitivity indices. An estimate of the total sensitivity indices is also included at no extra cost. The sampling method generates orthogonal samples.

7.4.9 How to induce dependencies in the input factors

The user can specify a dependency structure in the sample design. SIMLAB implements three methods to induce a dependency structure.

The Iman and Conover method

This is used to induce a desired *rank correlation* on pairs of input factors. Its characteristics are:

- rank correlations can be set independently on marginal distributions;

- the original form of the marginal distributions is preserved;

- it may be used with many sample schemes;

- if the correlations imposed are ill-defined, the resulting rank correlation matrix may not be positive definite, and an error message is displayed.

The dependence-tree/copula method

This method is related to influence diagrams, but makes use of undirected acyclic graphs instead of directed graphs that are used in influence diagrams. The user can specify correlations between input factors that form a tree structure. Whatever correlation values are imposed by the user in this way, it is guaranteed that a joint pdf exists. The joint pdf has the minimum information amongst all those joint distributions that satisfy the criteria given by the user.

The Stein method

This method allows the user to generate an LHS sample from any type of non-orthogonal sample. The user must provide an ASCII file (see format in Appendix C of the manual) that contains a non-orthogonal sample (e.g., a random sample, or even an empirical sample generated by an experiment). The method generates an LHS sample with the same dependency structure of the sample provided by the user.

7.5 How to execute models

There are different modalities to execute a model with SIMLAB.

In the external modality the users use a stand-alone application in which they have coded the model. Examples of such applications are the executable files that the users can find in .*SIMLAB**models*.

The application is built so as to read the sample file and to produce an output file using the format specifications of SIMLAB (see Appendix C in the on-line manual). The external model execution was studied to allow the interface to complex models.

SIMLAB can also run models built in Microsoft Excel^{TM}. SIMLAB generates the sample and saves all the information in an Excel worksheet called 'Inputs'. The model in Excel must be built so as to read from this worksheet the inputs one row at a time and to write the model outputs into another worksheet called 'Outputs'. When the model runs have terminated, Simlab imports the model outputs from the Worksheet 'Outputs' and the sensitivity calculations can start. Time dependent model outputs cannot be used in the Excel environment.

It is also possible to perform sensitivity analyses in batch mode. First, the users generate the sample with SIMLAB. Then they call an external model to execute the runs, and, finally, they supply SIMLAB with a model output file.

In the internal modality, a parser helps the user to edit straightforward equations and build simple analytical models (no time dependent outputs). This allows the user to conduct quick and easy tests.

7.6   Sensitivity analysis

With the previous stage the user has created a mapping from input factors to output results of the form

$$[y_1^{(i)}, y_2^{(i)}, \ldots, y_m^{(i)}, x_1^{(i)}, x_2^{(i)}, \ldots, x_k^{(i)}], i = 1, \ldots, N$$

where m is the number of model outputs, k is the number of model inputs and N is the number of model evaluations.

This mapping can be explored in many ways to determine the sensitivity of model predictions to individual input variables.

The right-hand frame of the SIMLAB main panel is dedicated to this task.

Uncertainty analysis is straightforward. Means, variances and distribution functions can be estimated directly from the model predictions. The user can set a number of parameters related to the calculation of some statistical hypothesis tests. Examples of tests implemented in SIMLAB are: the Smirnov test, used to test the hypothesis that a sample comes from a particular distribution; and the Tchebycheff's and t-tests to compute confidence bounds on μ, the population mean of Y.

For sensitivity analysis, a number of techniques are available in SIMLAB. The generation of *scatter plots* is undoubtedly the simplest sensitivity analysis technique. This approach consists of generating plots of the points $(x_j^{(i)}, y^{(i)})$, $i = 1, \ldots, N$, for each independent variable X_j. Scatter plots offer a qualitative measure of sensitivity.

Another simple measure of sensitivity is the *Pearson product moment correlation coefficient (PEAR)* which is the usual linear correlation coefficient computed on the

$$x_j^{(i)}, y^{(i)} (i = 1, \ldots, N).$$

For non-linear models the *Spearman coefficient (SPEA)* is preferred as a measure of correlation. This is essentially the same as PEAR, but uses the ranks of both Y and X_j instead of the raw values i.e.,

$$\text{SPEA}(Y, X_j) = \text{PEAR}(R(Y), R(X_j))$$

where R(.) indicates the transformation that substitutes the variable value with its rank.

More quantitative measures of sensitivity are based on regression analysis. *Standardised Regression Coefficients (SRC)* quantify the linear effect of each input variable (see Box 2.2 in Chapter 2).

The Partial Correlation Coefficient (PCC) gives the strength of the correlation between Y and a given input X_j cleaned of any effect due to correlation between X_j and any other input. In other words, PCC provides a measure of variable importance that tends to exclude the effects of other variables. In the particular case in which the input variables are orthogonal, the order of variable importance based on either SRC or PCC (in their absolute values) is exactly the same.

Regression analysis often performs poorly when the relationships between the input variables are non-linear. The problem associated with poor linear fits to non-linear data can often be avoided with the use of the rank transformations. *Standardised*

Rank Regression Coefficients are the SRC calculated in terms of $R(y_i)$ and $R(x_k)$.

Similarly, the PCC can be computed on the ranks giving the *Partial Rank Correlation Coefficients*.

The extension of the method of Morris presented in this book estimates two indicators (μ^* and σ). Small values for μ^* point to factors with negligible effect; σ measures the strength of the interaction effects. This method can be used only if the Morris sampling plan has been selected (Chapter 4).

The *FAST method* produces model-free estimates of first-order sensitivity indices using the algorithm proposed by Cukier *et al.* (1973). An extension of the FAST method is implemented in SIMLAB, which produces estimates of both first-order indices and total effect indices. SIMLAB also implements the extended FAST for groups of factors. SIMLAB provides visualisation of the indices in the form of pie-charts. This method can be used only if the corresponding FAST sampling plan has been selected.

The importance measure is model-free sensitivity technique that supplies the first-order indices (main effects) of individual factors. This measure is less efficient than others for orthogonal input. It becomes useful when the inputs are non-orthogonal. The importance measure uses r-LHS as sampling design.

The method of Sobol' produces model-free estimates of first-order sensitivity indices, higher-order indices and total indices using the algorithm proposed by Sobol' 1990. SIMLAB provides visualisation of the indices in the form of pie-charts. This method can be used only if the corresponding Sobol' sampling plan has been selected.

8 FAMOUS QUOTES: SENSITIVITY ANALYSIS IN THE SCIENTIFIC DISCOURSE

Sensitivity analysis is considered by some as a prerequisite for model building in any setting, be it diagnostic or prognostic, and in any field where models are used. Kolb, quoted in (Rabitz 1989), noted that theoretical methods are sufficiently advanced, so that it is intellectually dishonest to perform modelling without SA. Fürbinger (1996) muses:

Sensitivity analysis for modellers?
Would you go to an orthopaedist who didn't use X-ray?

Among the reasons for an increased role of sensitivity analysis in the scientific discourse is the change in the role of science in society in the last decade. This has seen the emergence of issues such as legitimacy (the end of scientists' purported neutrality; the need to cope with plurality of frames of reference and value judgement . . .) and relevance (models are questioned). Quantitative sensitivity analysis becomes a prescription in this context, as part of the quality assurance of the process. Especially crucial is its role in contributing to the defensibility of model-based analysis. Some quotes from practitioners make the point.

According to Hornberger and Spear (1981):

. . . most simulation models will be complex, with many parameters, state-variables and non linear relations. Under the best circumstances, such models have many degrees of freedom and, with judicious fiddling, can be made to produce virtually any desired behaviour, often with both plausible structure and parameter values.

Examples of instrumental use of models can be found in the literature, especially when models are used for making decisions that

Sensitivity Analysis in Practice: A Guide to Assessing Scientific Models A. Saltelli, S. Tarantola,
F. Campolongo and M. Ratto © 2004 John Wiley & Sons, Ltd.   ISBN 0-470-87093-1

will have a large social and economic impact. Thus, it is not surprising to meet cynical opinions about models. An example was in *The Economist* (1998) where one reads that:

based largely on an economic model . . . completing K2R4 [a nuclear reactor] in 2002 has a 50% chance of being 'least cost'.

Given that the model was used to contradict a panel of experts on the opportunity to build the aforementioned reactor, *The Economist* comments:

Cynics say that models can be made to conclude anything provided that suitable assumptions are fed into them.

The problem, highlighted by Hornberger and illustrated by the example above, is acutely felt in the modelling community. An economist, Edward E. Leamer (1990), has a solution:

I have proposed a form of organised sensitivity analysis that I call 'global sensitivity analysis' in which a neighbourhood of alternative assumptions is selected and the corresponding interval of inferences is identified. Conclusions are judged to be sturdy only if the neighbourhood of assumptions is wide enough to be credible and the corresponding interval of inferences is narrow enough to be useful.

This awareness of the dangers implicit in selecting a model structure as true and working happily thereafter leads naturally to the attempt to map rigorously alternative model structures or working hypotheses into the space of the model predictions. The natural extension of this is the analysis of how much each source of uncertainty weights on the model prediction. One possible way to apportion the importance of the input factor with respect to the model output is to apply global quantitative sensitivity analysis methods. Here the expression, 'Global Sensitivity Analysis', takes on an additional meaning, with respect to that proposed by Leamer, in that a decomposition of the total uncertainty is sought.

Hornberger's concern about models is better known in the scientific community as the problem of the GIGO models (Garbage In-Garbage Out).[1] There is apparently even an operative definition of a GIGO principle (Stirling, 1998):

---

[1] Assuming one has already got rid of the garbage in between, i.e. numerical or conceptual code errors.

Precision of outputs goes up as accuracy of inputs goes down.

In other words, one way of GIGOing is to obtain precise outputs by arbitrarily restricting the input space.

Andrew Stirling studies 'precautionary' and 'science based' approaches to risk assessment and environmental appraisal. In a recent work, which is the compilation of four different studies on the subject, he studies what the precautionary principle implies and how can it be operationalised (Stirling, 1999a,b). One of the recommendations he arrives at is 'Express Analytical Results Using Sensitivity Analysis' (Stirling, 1999b, p. 78):

It has been shown in this interim report that – in a variety of areas – risk assessment results are often presented with a very fine degree of numerical precision. Such a style conveys the impression of great accuracy, and distracts attention from the crucial question of the sensitivity of final results to changes in starting assumptions. This problem is particularly acute, where the values obtained – and even the ordering of different options – are quite volatile under the perspectives in appraisal associated with different social constituencies and economic interests. A practical and well-established way of dealing with such a problem lies in 'sensitivity analysis' – a technique involving the explicit linking of alternative framing assumptions with the results, which they yield. Rather than being expressed as discrete scalar numbers, then, risk assessment results might be expressed as ranges of values, with the ends of the ranges reflecting extremities in the framing assumptions associated with different stakeholders in the appraisal process.

Stirling introduces in this text the value-laden nature of different framing assumptions, a familiar topic in present-day discourse on governance (see also Lemons *et al.*, 1997).

One more illustration of how sensitivity analysis (or the lack of it) might impinge on the defensibility of a model-based analysis is the following.

A team led by Daniel Esty of Yale University, with support from Columbia University, produced on behalf of the World Economic Forum a new Environmental Sustainability Index (ESI, 2001), and presented it to the annual Davos summit in 2001. This study contains a detailed assessment of dozens of variables that influence the environmental health of economies, producing an overall index that allows countries to be ranked. Mathis Wackernagel, intellectual father of the 'Ecological Footprint' and thus an authoritative source in the sustainable development community, concludes a critique of the study done by Daniel Esty *et al.* by noting (Wackernagel, 2001):

Overall, the report would gain from a more extensive peer review and a sensitivity analysis. The lacking sensitivity analysis undermines the confidence in the results since small changes in the index architecture or the weighting could dramatically alter the ranking of the nations.

It is clear from this example that index numbers, such as the ESI, can be considered as models. In Saltelli *et al.* (2000a, p. 385) it has been shown how SA can be used to put an environmental debate on track by suggesting that the uncertainty in the decision on whether to burn or otherwise dispose of solid urban waste depends on the choice of the index and not on the quality of the available data (e.g. emission factors).

Oreskes *et al.* (1994) in an article in *Science* entitled 'Verification, validation and confirmation of numerical models in the earth sciences', puts SA in an apparently different context. The SA is not treated as a tool to build or improve a model, but it represents one of the possible licit uses that can be done of the model itself. According to Oreskes, who takes a Popperian stance on the issue, natural systems are never closed, and models put forward as description of these are never unique. Hence, models can never be 'verified' or 'validated', but only 'confirmed' or 'corroborated' by the demonstration of agreement (non-contradiction) between observation and prediction. Since confirmation is inherently partial, models are qualified by a heuristic value: models are representations, useful for guiding further study, but not susceptible to proof. In Oreskes *et al.*'s point of view:

Models can corroborate a hypothesis...Models can elucidate discrepancies with other models. Models can be used for sensitivity analysis – for exploring 'what if' questions – thereby illuminating which aspects of the system are most in need of further study, and where more empirical data are most needed.

A last quote for this chapter is from Peter Høeg, a Danish novelist, who notes in his excellent *Borderliners* (1995):

That is what we meant by science. That both question and answer are tied up with uncertainty, and that they are painful. But that there is no way around them. And that you hide nothing; instead, everything is brought out into the open.

Høeg, like Oreskes, seems to think that uncertainty is not an accident of the scientific method, but its substance.

Summing up the thoughts collected so far, one could say that the role of scientists in society today is not that of revealing truth, but rather of providing evidence, be it 'crisp' or circumstantial, based on incomplete knowledge, sometimes in the form of probability, before and within systems of conflicting stakes and beliefs (see also Funtowicz *et al.*, 1996).

As a result, scientists need to provide evidence that is defensible, transparent in its assumptions and comparable against and across different framing assumptions. The term 'socially robust knowledge' is also used by some to identify that process whereby different views of the issues, different value judgement systems and framing assumptions, have been incorporated into the analysis. In most controversial debates where science plays a role, the negotiation takes place in the space of the uncertainties that arise both from the poor understanding of the issue and the different expectations and values referred to it. Characterising the uncertainties is an essential ingredient of the process, and this entails sensitivity analysis.

REFERENCES

Adriaanse, A. (1993) *Environmental Policy Performance Indicators*. Sdu Uitgeverij Koninginnegracht, The Hague.

Andres, T. H. and Hajas, W. C. (1993) Using iterated fractional factorial design to screen parameters in sensitivity analysis of a probabilistic risk assessment model. In: *Proceedings of the Joint International Conference on Mathematical Methods and Supercomputing in Nuclear Applications, Karlsruhe, Germany, 19–23 April 1993* (eds. H. Küsters, E. Stein and W. Werner) Vol. 2, 328–337.

Bard, Y. (1974) *Nonlinear parameter estimation*. Academic Press, New York.

Beck, M. B. (1987) Water quality modelling: a review of the analysis of uncertainty. *Water Resources Research*. **23**, 1393–1442.

Bettonvil, B. (1990) *Detection of important factors by sequential bifurcation*. Tilburg University Press, Tilburg.

Bettonvil, B. and Kleijnen, J. P. C. (1997) Searching for important factors in simulation models with many factors: sequential bifurcation. *Eur. J. Oper. Res.* **96**(1), 180–194.

Beven, K. J. and Binley, A. M. (1992) The future of distributed models: model calibration and uncertainty prediction. *Hydrol. Process.* **6**, 279–298.

Beven, K. J. (2001) *Rainfall-Runoff Modelling: The Primer*. John Wiley & Sons, Ltd, Chichester.

Beven, K. J., Freer, J., Hankin, B. and Schulz, K. (2000) The use of generalised likelihood measures for uncertainty estimation in high order models of environmental systems. In: *Nonlinear and Nonstationary Signal Processing* (eds W. J. Fitzgerald, R. L. Smith, A. T. Walden and P. C. Young), Cambridge University Press, 115–151.

Box, G. E. P., Hunter, W. J. and Hunter, J. S. (1978) *Statistics for Experimenters. An Introduction to Design, Data Analysis and Model Building*. John Wiley & Sons, Ltd, New York.

Campolongo, F., Cariboni, J. and Saltelli, A. (2003) In progress.

Campolongo, F. and Saltelli, A. (1997) Sensitivity analysis of an environmental model; an application of different analysis methods. *Reliab. Engng. Syst. Safety* **57**(1), 49–69.

Campolongo, F., Tarantola, S. and Saltelli, A. (1999) Tackling quantitatively large dimensionality problems. *Comput. Phys. Commun.* **117**, 75–85.

Sensitivity Analysis in Practice: A Guide to Assessing Scientific Models A. Saltelli, S. Tarantola, F. Campolongo and M. Ratto © 2004 John Wiley & Sons, Ltd. ISBN 0-470-87093-1

Campolongo, F. and Rossi, A. (2002) Sensitivity analysis and the delta hedging problem. In: *Proceedings of PSAM6, 6th International Conference on Probabilistic Safety Assessment and Management, Puerto Rico, June 23–28 2002.*

Caswell, H. (1989) *Matrix Populations Models: Construction, Analysis and Interpretation.* Sinauer, Sunderland, Massachusetts.

Cotter, S. C. (1979) A screening design for factorial experiments with interactions. *Biometrika* 66(2), 317–320.

CPC (1999) Special issue on sensitivity analysis. *Comput. Phys. Commun.* 117, 1–2.

Crank, J. (1975) *The Mathematics of Diffusion.* Oxford University Press, New York, 2nd edn.

Crosetto, M. and Tarantola, S. (2001) Uncertainty and sensitivity analysis: tools for GIS-based model implementation. *Int. J. Geogr. Inf. Sci.* 15(5), 415–437.

Cukier, R. I., Fortuin, C. M., Schuler, K. E., Petschek, A. G. and Schaibly, J. H. (1973) Study of the sensitivity of coupled reaction systems to uncertainties in rate coefficients. I Theory. *J. Chem. Phy.* 59(8), 3873–3878.

Cukier, R. I., Levine, H. B. and Schuler, K. E. (1978) Nonlinear sensitivity analysis of multiparameter model systems. *J. Comput. Phy.* 26, 1–42.

Draper, N. R. and Smith, H. (1981) *Applied Regression Analysis,* John Wiley & Sons, Ltd, New York.

EC (2002) European Commission Communication on Extended Impact Assessment. Brussels, 05/06/2002 COM(2002) 276 final. Guidelines for implementing the directive are available at the Governance page of the EC http://europa.eu.int/comm/governance/docs/index_en.htm.

EPA (1999) The US Environmental Protection Agency Science Policy Council, White Paper on the Nature and Scope of Issues on Adoption of Model Use Acceptability Guidance, http://www.epa.gov/osp/crem/library/whitepaper_1999.pdf.

ESI (Environmental Sustainability Index) (2001) http://www.ciesin.org/indicators/ESI/index.html, INTERNET.

Freer, J., Beven, K. and Ambroise, B. (1996) Bayesian estimation of uncertainty in runoff prediction and the value of data: an application of the GLUE approach. *Water Resources Research* 32(7), 2161–2173.

Funtowicz, S., O'Connor, M., Faucheux, S., Froger, G. and Munda, G. (1996) Emergent complexity and procedural rationality: post-normal science for sustainability. In: *Getting Down to Earth* (eds R. Costanza, S. Olman and J. Martinez-Alier).

Fürbinger, J. M. (1996) Sensitivity analysis for modellers. *Air Infiltration Review* 17(4).

Geweke, J. (1999) Using simulation methods for Bayesian econometric models: inference, development and communication (with discussion and rejoinder). *Economet. Rev.* 18, 1–126.

Grieb, T. M., Shang, N., Spear, R. C., Gherini, S. A. and Goldstein, R. A. (1999) Examination of model uncertainty and parameter interaction in the global carbon cycling model. *Environ. Int.* **25**, 787–803.

Griewank, A. (2000) *Evaluating derivatives, Principles and techniques of algorithmic differentiation.* SIAM Publisher, Philadelphia.

Hammersly, J. M. and Handscomb, D. C. (1964) *Monte Carlo Methods.* Methuen and Company, London.

Helton, J. C. (1993) Uncertainty and sensitivity analysis techniques for use in performance assessment for radioactive waste disposal. *Reliab. Engng Syst. Safety* **42**, 327–367.

Henrion, M. and Fischhoff, B. (1986) Assessing uncertainty in physical constants. *Am. J. Phys.* **54**(9), 791–799.

Hoeting, J. A., Madigan, D., Raftery, A. E. and Volinsky, C. T. (1999) Bayesian model averaging: a tutorial. *Statist. Sci.* **14**, 382–417.

Homma, T. and Saltelli, A. (1996) Importance measures in global sensitivity analysis of model output. *Reliab. Engng Syst. Safety* **52**(1), 1–17.

Hornberger, G. M. and Spear, R. C. (1981) An approach to the preliminary analysis of environmental systems. *J. Environ. Manage.* **12**, 7–18.

Høeg, P. (1995) *Borderliners.* McClelland-Bantam, Toronto. p. 19.

Iman, R. L. and Conover, W. J. (1982) A distribution free approach to inducing rank correlation among input variables. *Commun. Stat. – Simul. C.* **11**(3), 311–334.

Iman, R. L. and Shortencarier, M. J. (1984) A Fortran 77 program and user's guide for the generation of Latin hypercube and random samples for use with computer models. SAND83-2365. Sandia National Laboratories, Albuquerque, NM.

Ishigami, T. and Homma, T. (1990) An importance quantification technique in uncertainty analysis for computer models. In: *Proceedings of the ISUMA '90. First International Symposium on Uncertainty Modelling and Analysis, University of Maryland, USA, December 3–5, 1990,* pp. 398–403.

Jansen, M. J. W., Rossing, W. A. H. and Daamen, R. A. (1994) Monte Carlo estimation of uncertainty contributions from several independent multivariate sources. In: *Predictability and Nonlinear Modelling in Natural Sciences and Economics* (eds J. Grasman and G. van Straten) Kluwer Academic Publishers, Dordrecht, pp. 334–343.

JSCS (1997) Special Issue on Sensitivity Analysis. *J. Statist. Comput. Simul.* **57**, 1–4.

Kass, R. E. and Raftery, A. E. (1995) Bayes Factors, *J. Am. Statist. Assoc.* **90**(430), 773–795.

Koda, M., McRae, G. J. and Seinfeld, J. H. (1979) Automatic sensitivity analysis of kinetic mechanisms. *Int. J. Chem. Kinet.* **11**, 427–444.

Krzykacz-Hausmann, B. (2001) Epistemic sensitivity analysis based on the concept of entropy. In: *Proceedings of SAMO2001* (eds P. Prado and R. Bolado) Madrid, CIEMAT, pp. 31–35.

Leamer, E. E. (1990) 'Let's take the con out of econometrics" and "Sensitivity Analysis would help". In: *Modelling Economic Series* (ed. C. W. J. Granger) Clarendon Press, Oxford.

Lemons, J., Shrader-Frechette, K. and Cranor, C. (1997) The precautionary principle: scientific uncertainty and type I and II errors. *Found. Sci.* **2**, 207–236.

McKay, M. D. (1995) Evaluating prediction uncertainty. Technical Report NUREG/CR-6311, LA-12915-MS, US Nuclear Regulatory Commission and Los Alamos National Laboratory.

McKay, M. D. (1996) Variance-based methods for assessing uncertainty importance in NUREG-1150 analyses. LA-UR-96-2695, Los Alamos National Laboratory.

McKay M. D., Beckman, R. J. and Conover, W. J. (1979) A comparison of three methods of selecting values of input variables in the analysis of output from a computer code. *Technometrics.* **21**(2), 239–245.

Morris, M. D. (1991) Factorial sampling plans for preliminary computational experiments. *Technometrics* **33**(2), 161–174.

OECD (1989) OECD/NEA PSAC User group, *PSACOIN Level E intercomparison* (eds B. W. Goodwin, J. M. Laurens, J. E. Sinclair, D. A. Galson and E. Sartori). Nuclear Energy Agency, Organisation for Economic Co-operation and Development, Paris.

OECD (1993) OECD/NEA PSAG User group, *PSACOIN Level S intercomparison* (eds A. Alonso, P. Robinson, E. J. Bonano and D. A. Galson), Nuclear Energy Agency, Organisation for Economic Cooperation and Development, Paris.

Oreskes, N., Shrader-Frechette, K. and Belitz, K. (1994) Verification, validation, and confirmation of numerical models in the earth sciences. *Science* **263**, 641–646.

Osidele, O. O. (2001) Reachable futures, structural change, and the practical credibility of environmental simulation models. PhD Thesis, University of Georgia, Athens.

Osidele, O. O. and Beck M. B. (2001) Identification of model structure for aquatic ecosystems using regionalized sensitivity analysis. *Water Sci. Technol.* **43**(7), 271–278.

Prado, P., Homma, T. and Saltelli, A. (1991) Radionuclide migration in the geosphere: a 1D advective and dispersive transport module for use in probabilistic system assessment codes. *Radioactive Waste Manag. Nucl. Fuel Cycle* **16**, 49–68.

Puolamaa, M., Kaplas, M. and Reinikainen, T. (1996) *Index of Environmental Friendliness – A Methodological Study*. Statistics Finland.

Rabitz, H. (1989) System analysis at molecular scale. *Science* **246**, 221–226.

Rabitz, H., Aliş Ö. F., Shorter, J. and Shim, K. (1999) Efficient input–output model representations. *Comput. Phys. Commun.* **117**, 11–20.

Ratto, M. (2003). In progress.

Ratto, M., Tarantola, S. and Saltelli, A. (2001) Sensitivity analysis in model calibration: GSA-GLUE approach. *Comput. Phys. Commun.* **136**, 212–224.

Ratto, M. and Tarantola, S. (2003) In progress.

Rebonato, R. (1998) *Interest-rate option models*, 2nd edn. John Wiley & Sons, Ltd, England.

RESS (1997) Special Issue on Sensitivity Analysis. *Reliab. Engng. Syst. Safety* **57**, 1.

RESS (2002) Special Issue: SAMO 2001: Methodological advances and innovative applications of sensitivity analysis. *Reliab. Engng. Syst. Safety* **79**, 2.

Ripley, R. D. (1987) *Stochastic Simulation.* John Wiley & Sons, Ltd, New York.

Risk Newsletter (1987) Defining Risk editorial note, *Risk Newsletter*, 7(3), 5.

Robinson, P. C. (2000) Personal Communication.

Robinson, P. C. and Hodgkinson, D. P. (1987) Exact solutions for radionuclide transport in the presence of parameter uncertainty. *Radioactive Waste Manag. Nucl. Fuel Cycle* 8(4), 283–311.

Romanowicz, R., Beven, K., Tawn, J. (1994) Evaluation of predictive uncertainty in nonlinear hydrological models using a Bayesian approach. In: *Statistics for the Environment 2, Water Related Issues* (eds V. Barnett and F. Turkman), John Wiley & Sons, Ltd, New York, pp. 297–315.

Romanowicz, R. and Beven, K. (1998) Dynamic real-time prediction of flood inundation probabilities. *Hydrol. Sci.* **43**(2), 181–196.

Romanowicz, R., Higson, H. and Teasdale, I. (2000) Bayesian uncertainty estimation methodology applied to air pollution modelling. *Environmetrics* **11**, 351–371.

Sacks, J., Welch, W. J., Mitchell, T. J. and Wynn, H. P. (1989) Design and analysis of computer experiments. *Statist. Sci.* **4**, 409–435.

Saisana, M. and Tarantola, S. (2002) State-of-the-art report on current methodologies and practices for composite indicator development, EUR 20408 EN.

Saltelli, A. (1999) Sensitivity analysis. Could better methods be used? *J. Geophys. Res.* **104**(D3), 3789–3793. Errata Corrige in **104**(D19), 24,013.

Saltelli, A. (2002) Making best use of model evaluations to compute sensitivity indices. *Comput. Phys. Commun.* **145**, 280–297.

Saltelli, A., Chan, K. and Scott, M. (Eds) (2000a) *Sensitivity Analysis.* John Wiley & Sons, Ltd, New York.

Saltelli, A. and Sobol', I. M. (1995). About the use of rank transformation in sensitivity analysis of model output. *Reliab. Eng. Syst. Safety* **50**(3), 225–239.

Saltelli, A. and Tarantola, S. (2002) On the relative importance of input factors in mathematical models: safety assessment for nuclear waste disposal. *J. Am. Statist. Assoc.* **97**(459), 702–709.

Saltelli, A., Tarantola, S. and Campolongo, F. (2000b) Sensitivity analysis as an ingredient of modelling. *Statist. Sci.* 15(4), 377–395.

Saltelli, A., Tarantola, S. and Chan, K. (1999) A quantitative, model independent method for global sensitivity analysis of model output. *Technometrics* 41(1), 39–56.

Sobol', I. M. (1990) Sensitivity estimates for nonlinear mathematical models. *Matematicheskoe Modelirovanie* 2, 112–118 (in Russian). [Transl. (1993) Sensitivity analysis for non-linear mathematical models. *Math. Modelling & Comp. Exp.* 1, 407–414.]

Spear, R. C. (1997) Large simulation models: calibration, uniqueness and goodness of fit. *Environ. Modell. Softw.* 12, 219–228.

Spear, R. C., Grieb, T. M. and Shang, N. (1994) Factor uncertainty and interaction in complex environmental models. *Water Resour. Res.* 30, 3159–3169.

Stein, M. (1987) Large sample properties of simulations using Latin hypercube sampling. *Technometrics.* 29(2), 143–151.

Stirling, A. (1998) Valuing the environmental impacts of electricity production: a critical review of some 'first generation' studies. *Energ. Source.* 20, 267–300.

Stirling, A. (1999a) On science and precaution in the management of technological risk, Volume I, A synthesis report of case studies. ESTO-IPTS, EUR 19056 EN.

Stirling, A. (1999b) On science and precaution in the management of technological risk, Volume II, Case studies. ESTO-IPTS, EUR 19056 EN/2.

The Economist (1998) More fall-out from Chernobyl, June 27th, p. 98.

Tarantola, S., Saisana, M., Saltelli, A., Schmiedel, F. and Leapman, N. (2002) Statistical techniques and participatory approaches for the composition of the European Internal market Index 1992–2001, EUR 20547 EN.

Turanyi, T. (1990) Sensitivity analysis of complex kinetic systems. Tools and applications. *J. Math. Chem.* 5, 203–248.

Young, P. C. (1999) Data-based mechanistic modelling, generalised sensitivity and dominant mode analysis. *Comput. Phys. Commun.* 117, 113–129.

Young, P. C., Parkinson, S. D. and Lees, M. (1996) Simplicity out of complexity: Occam's razor revisited. *J. Appl. Stat.* 23, 165–210.

Young, P. C. and Romanowicz, R. (2003) A Comment on GLUE, Centre for Research on Environmental Systems and Statistics, Lancaster University, U.K., CRES Technical Report Number TR/180.

Wackernagel, M. (2001) Shortcomings of the Environmental Sustainability Index. Redefining Progress, 6 March, http://www.rprogress.org/about/.

Zaldivar, J. M., Kourti, N., Villacastin, C., Strozzi, F. and Campolongo, F. (1998) Analysing dynamics of complex ecologies from natural recordings: an application to fish population models. JRC-ISIS Technical Note No. I.98.199.

INDEX

Printed and bound by CPI Group (UK) Ltd, Croydon, CR0 4YY

27/10/2024

14580285-0002